Mark Girolami (Ed.)

Advances in Independent Component Analysis

Springer

Mark Girolami, BSc (Hons), BA, MSc, PhD, CEng, MIEE, MIMechE
Department of Computing and Information Systems, University of Paisley,
High Street, Paisley, PA1 2BE

Series Editor
J.G. Taylor, BA, BSc, MA, PhD, FInstP
Centre for Neural Networks, Department of Mathematics, King's College,
Strand, London WC2R 2LS, UK

ISSN 1431-6854

ISBN 1-85233-263-8 Springer-Verlag London Berlin Heidelberg

British Library Cataloguing in Publication Data
Advances in independent component analysis. - (Perspectives
  In neural computing)
  1. neural networks (computer science)
  I. Girolami, Mark, 1963-
  006.3'2
  ISBN 1852332638

Library of Congress Cataloging-in-Publication Data
Advances in independent component analysis / Mark Girolami (Ed.).
      p. cm. -- (Perspectives in neural computing)
    Includes bibliographical references and index.
    ISBN 1-85233-263-8 (alk. paper)
      1. Neural networks (Computer science) 2. Multivariate analysis. I. Girolami, Mark,
  1963- II. Series.
  QA76.87. A378 2000-05-05
  006.3'2--dc21   00-037370

Typesetting: Camera ready by editor.
Printed and bound by the Athenæum Press Ltd., Gateshead, Tyne & Wear
34/3830-543210 Printed on acid-free paper SPIN 10755275

Perspectives in Neural Computing

Springer
London
Berlin
Heidelberg
New York
Barcelona
Hong Kong
Milan
Paris
Singapore
Tokyo

Preface

Since 1995 the neural computing research community has produced a large number of publications dedicated to the Blind Separation of Sources (BSS) or Independent Component Analysis (ICA). This book is the outcome of a workshop which was held after the 1999 International Conference on Artificial Neural Networks (ICANN) in Edinburgh, Scotland. Some of the most active and productive neural computing researchers gathered to present and share new ideas, models and experimental results and this volume documents the individual presentations. It is satisfying to see that some of the more *difficult* extensions of the basic ICA model such as nonlinear and temporal mixing have been considered and the proposed methods are presented herein. It is also abundantly clear that ICA has been embraced by a number of researchers involved in biomedical signal processing as a powerful tool which in many applications has supplanted decomposition methods such as Singular Value Decomposition (SVD). No less than six of the chapters in this book consider, directly or indirectly, biomedical applications of ICA.

The first section of this book deals with temporal aspects which can sometimes be associated with ICA modelling. In Chapter 1, Penny *et al.* propose an ICA-based generative model for the analysis of multivariate time series. The model itself is a combination of Hidden Markov Models (HMMs), ICA and Generalised Autoregressive (GAR) models. The estimation of the parameters for each component are estimated in a comprehensive Expectation Maximisation (EM) framework. The overall model is shown to be able to learn the dynamics of the underlying sources, learn the parameters of the mixing process and detect discrete changes in either the source dynamics or the mixing or both. The overall model can then partition a multivariate time series into different regimes and so may be used for classification purposes. The main application of this model is for biomedical signal processing and, in particular, analysis of the electroencephalogram (EEG). Other potential applications may include multivariate financial time series analysis and machine condition monitoring, indeed any application which exhibits strong non-stationary generative processes.

In the second chapter, Everson *et al.* propose a method for performing ICA on time varying mixtures where the mixing matrix is no longer a stationary set of parameters to be estimated but a set which varies in time and therefore requires their evolution to be consistently tracked. As this is a nonlinear and non-Gaussian tracking problem, the Kalman filter, whose observation equation is linear and state densities are Gaussian, is inappropriate for this task and so the state density is estimated using a sampling method. For this, Everson *et al.* propose the use of particle filters which represent the state density by a swarm of "particles". This provides an elegant method of estimating and tracking the mixing parameters in a non-stationary manner. This concludes the first section of the book which has considered temporal effects on the basic ICA model.

The second section contains two chapters which consider the very core of ICA, the independence assumption. In Chapter 3 Hyvärinen *et al.* propose the notion of topographic ICA. This is a generative model that combines topographic mapping with the linear ICA model. In the proposed topographic ICA, the distance between represented components is defined by the mutual information implied by the higher-order correlations, which gives the natural distance measure in the context of ICA. Most topographic mapping methods define the topographical distance as the geometrical Euclidean distance. What is most interesting about this approach is that it is possible to define a topography among a set of orthogonal vectors, whose Euclidean distances are all equal. A simple gradient-based form of Hebbian learning is developed to estimate the model parameters. The experiments detailed to demonstrate the utility of topographic ICA include a most impressive biomedical application which indicates the promise of this method as a further tool for multivariate data analysis. The second contribution in this section by Barros extends the ICA framework to include what is termed non-orthogonal signals. The assumption of statistical independence among the sources is discarded and a strategy is suggested for decreasing redundancy, in the sense that the output signals should have minimal mutual spectral overlap. To differentiate it from ICA, this approach is referred to as *dependent component analysis* (DCA). An experiment shows the separation ability of the proposed method when applied to cyclo-stationary signals.

The third section deals with the complex problems of nonlinear ICA and the enhancement of degraded images by blind deconvolution and separation. As a precursor to these a tutorial chapter regarding Ensemble Learning is provided by Lappalainen and Miskin. For many problems, it is intractable to perform inferences using the true posterior density over the unknown variables. Ensemble Learning allows the true posterior to be approximated by a simpler approximate distribution for which the required inferences therefore become tractable. The sixth chapter deals with the problem of the nonlinear counterparts of PCA and ICA where the generative mapping from original sources to observed data is not restricted to being linear. The general nonlinear mapping from sources to data observations (corrupted by Gaussian noise) is modeled by a multi-layer perceptron network and the distributions of source signals are modeled by mixture-of-Gaussians. The nonlinear mapping, the source distributions and the noise level are estimated from the data. In employing the Bayesian approach to learning, Lappalainen and Honkela avoid problems with overlearning which would otherwise be severe in the unsupervised learning of such flexible nonlinear models. Miskin and MacKay utilise ensemble learning for the blind source separation and deconvolution of images. This of course finds application in the enhancement of images generated by telescopes observing the sky or medical imaging processes. The authors show that the use of ensemble learning allows the number of unobserved images to be inferred by maximising the bound on the evidence with respect to the number of hidden images. This then overcomes the problem with Maximum Likelihood methods which find only point estimates for the latent variables and consequently may suffer from over-fitting to the data.

The final section of this volume includes seven chapters which can broadly be described as dealing with the problem of data analysis. Chapter 8 sees Palmieri and Budillon address the problem of separating multi-class sources which have either rank-deficient distributions or show very concentrated eigenvalues. They approach the problem by assuming the observed samples from each class are confined to different linear subspaces. The criterion for optimisation is based on estimating an on-line membership function for each data point which combines a measure of the likelihood and the distance from the subspaces. The independent components are then searched for within each subspace. This particular method shows promise for applications such as feature extraction.

The next four chapters all deal with some form of biomedical signal processing, in particular data generated by observing neurobiological functions. In Chapter 9, Hansen develops an ICA method for high-dimensional noisy mixtures where the number of sources is much smaller than the number of sensors (pixels). Noting that the signal model under consideration has a distribution which takes the form of a hidden variable Gibbs distribution, the mixing matrix and the noise parameters can be inferred using the developed Boltzmann learning rule. The proposed method is successfully used for the analysis of a functional Magnetic Resonance Imaging sequence. Following on from this chapter, Vigário et al. provide a comprehensive review of recent applications of the linear ICA model to the analysis of biomagnetic brain signals. They show that ICA is able to differentiate between somatosensory and auditory brain responses in the case of vibrotactile stimulation. They also show that the extracted independent components also exhibit field patterns that agree with the conventional current dipole models. In addition they show that the application of ICA to an averaged auditory evoked response can isolate the main response, with a latency of about 100 ms, from subsequent components. As noted by the authors, ICA may indeed facilitate the understanding of the functioning of the human brain, as a finer mapping of the brain's responses may be achieved.

In Chapter 11, Ikeda considers two problems in analysing biological data, the amount of noise in the data, and the estimation of the number of components. It is argued that the standard approach (pre-whitening by PCA and further rotation by an orthogonal matrix) does not work well when the data are noisy and the number of the sources is unknown. Instead of whitening with PCA, a factor analysis model for preprocessing is proposed. An estimate of the number of the sources and the amount of the noise in the data is estimated using factor analysis. The proposed method is shown to be very effective by a substantial experiment with MEG data. Chapter 12, presents techniques for the recovery of component spatial and temporal modes from spatio-temporal data sets from medical imaging. These data sets were derived by direct optical imaging of the rat barrel cortex, before, during, and after whisker stimulation. Functional magnetic resonance imaging and optical imaging were used throughout. These techniques were developed in order to help address some current issues involving the nature of the haemodynamic response to neural activity.

The penultimate chapter provides details of a novel application of ICA: automatic content-based classification of text documents. Linear ICA is found to identify a generalizable low-dimensional basis set for the typically high-dimensional and noisy data representation of vector space model for text documents. The authors claim that the major benefit of using over Latent Semantic Analysis is that the ICA representation is better aligned with the content group structure. Two example collections (document corpora in information retrieval parlance) are used in the experimental work of this chapter with mixed results reported. The final chapter concludes this volume with a review of a number of biologically inspired and plausible independence seeking artificial neural network models.

To complement this volume a website which contains software, data and various useful links to many of the ICA based methods featured herein is available at the address below.

http://www.dcs.ex.ac.uk/ica/

Mark Girolami
University of Paisley, Scotland
February 2000

Contents

Part II   The Validity of the Independence Assumption

3   The Independence Assumption: Analyzing the Independence of the Components by Topography
Aapo Hyvärinen, Patrik O. Hoyer and Mika Inki 45

4   The Independence Assumption: Dependent Component Analysis
Allan Kardec Barros 63

Contributors

William D. Penny
Department of Engineering Science
Oxford University
Parks Road
Oxford, OX1 3PJ
United Kingdom
wpenny@robots.ox.ac.uk

Stephen J. Roberts
Department of Engineering Science
Oxford University
Parks Road
Oxford, OX1 3PJ
United Kingdom
sjrob@robots.ox.ac.uk

Richard M. Everson
Department of Computer Science
University of Exeter
Prince of Wales Road
Exeter, EX4 4PT
United Kingdom
R.M.Everson@exeter.ac.uk

Aapo Hyvärinen
Neural Networks Research Centre
Helsinki University of Technology
P.O. Box 5400
FIN-02015 HUT
Finland
aapo@james.hut.fi

Patrik O. Hoyer
Neural Networks Research Centre
Helsinki University of Technology
P.O. Box 5400
FIN-02015 HUT
Finland
Patrik.Hoyer@hut.fi

Mika Inki
Neural Networks Research Centre
Helsinki University of Technology
P.O. Box 5400
FIN-02015 HUT
Finland
inki@mail.cis.hut.fi

Allan Kardec Barros
Laboratory for Bio-Mimetic Systems
RIKEN
2271-130 Anagahora, Shimoshidami
Nagoya 463-0003
Japan
allan@bmc.riken.go.jp

Harri Lappalainen
Neural Networks Research Centre
Helsinki University of Technology
P.O. Box 5400
FIN-02015 HUT
Finland
Harri.Lappalainen@hut.fi

James W. Miskin
Cavendish Laboratory
University of Cambridge
Madingley Road
Cambridge, CB3 0HE
United Kingdom
jwm1003@mrao.cam.ac.uk

David J. C. MacKay
Cavendish Laboratory
University of Cambridge
Madingley Road
Cambridge, CB3 0HE
United Kingdom
mackay@mrao.cam.ac.uk

Lars Kai Hansen
Dept of Mathematical Modeling
Technical University of Denmark
DK-2800
Lyngby
Denmark
lkhansen@imm.dtu.dk

Francesco Palmieri
Dip. di Ing.
Elettronica e delle Telecomunicazioni
Università di Napoli
Federico II
Italy
frapalmi@unina.it

Alessandra Budillon
Dip. di Ing.
Elettronica e delle Telecomunicazioni
Università di Napoli
Federico II
Italy
alebudil@unina.it

Ricardo Vigário
Neural Networks Research Centre
Helsinki University of Technology
P.O. Box 5400
FIN-02015 HUT
Finland
rvigario@pyramid.hut.fi

Jaakko Särelä
Neural Networks Research Centre
Helsinki University of Technology
P.O. Box 5400
FIN-02015 HUT
Finland
Jaakko.Sarela@hut.fi

Erkki Oja
Neural Networks Research Centre
Helsinki University of Technology
P.O. Box 5400
FIN-02015 HUT
Finland
Erkki.Oja@hut.fi

Antti Honkela
Neural Networks Research Centre
Helsinki University of Technology
P.O. Box 5400
FIN-02015 HUT
Finland
ahonkela@james.hut.fi

Shiro Ikeda
Brain Science Institute
RIKEN
Hirosawa 2-1
Wako-Shi, Saitama, 351-0198
Japan
Shiro.Ikeda@brain.riken.go.jp

John Porrill
Department of Psychology
The University of Sheffield
Sheffield
S10 2TP
United Kingdom
J.Porrill@sheffield.ac.uk

James V. Stone
Department of Psychology
The University of Sheffield
Sheffield
S10 2TP
United Kingdom
J.V.Stone@sheffield.ac.uk

Peter Coffey
Department of Psychology
The University of Sheffield
Sheffield
S10 2TP
United Kingdom
P.Coffey@sheffield.ac.uk

John Mayhew
Department of Psychology
The University of Sheffield
Sheffield
S10 2TP
United Kingdom
J.E.Mayhew@sheffield.ac.uk

Jason Berwick
Department of Psychology
The University of Sheffield
Sheffield
S10 2TP
United Kingdom
PCP95JB@sheffield.ac.uk

Thomas Kolenda
Dept of Mathematical Modeling
Technical University of Denmark
DK-2800
Lyngby
Denmark
thko@imm.dtu.dk

Sigurdur Sigurdsson
Dept of Mathematical Modeling
Technical University of Denmark
DK-2800
Lyngby
Denmark
siggi@imm.dtu.dk

Pei Ling Lai
CIS Department
University of Paisley
High Street
Paisley, PA1 2BE
Scotland, United Kingdom
lai-ci0@wpmail.paisley.ac.uk

Darryl Charles
CIS Department
University of Paisley
High Street
Paisley, PA1 2BE
Scotland, United Kingdom
char-ci0@wpmail.paisley.ac.uk

Colin Fyfe
CIS Department
University of Paisley
High Street
Paisley, PA1 2BE
Scotland, United Kingdom
fyfe-ci0@wpmail.paisley.ac.uk

Foreword

The 9th International Conference on Artificial Neural Networks (ICANN 99) was held at the University of Edinburgh from 7-10 September 1999. On the following day four workshops were organized on the topics of: Interactions between Theoretical and Experimental Approaches in Developmental Neuroscience; Emergent Neural Computation Architectures Based on Neuroscience; Kernel Methods: Gaussian Process and Support Vector Machine Predictors; and Developments in Artificial Neural Network Theory: Independent Component Analysis and Blind Source Separation, the last of which gave rise to this volume. The holding of workshops was an innovation for ICANN, importing a feature that is popular at many North American conferences, including the Neural Information Processing Systems (NIPS) conference.

We would like to thank all of the workshop organizers for their hard work in putting together such high quality workshops that attracted over 100 attendees, and the workshop helpers for making the logistics run smoothly. The workshops were sponsored by the Division of Informatics (University of Edinburgh), the Institute for Adaptive and Neural Computation (University of Edinburgh), and the British Neuroscience Association, and support for the Emergent Neural Computation Architectures Based on Neuroscience workshop came from EPSRC (UK).

The workshop on Independent Components Analysis (ICA) was organized by Professor Mark Girolami. It attracted many of the leading researchers in the area, giving rise to a very strong programme, and we are delighted that the workshop led to this volume so that the advances discussed at the workshop can be shared with a wider audience.

Chris Williams
(Workshops Organizer)
David Willshaw and Alan Murray
(ICANN'99 Co-Chairs)
February 2000

Part I

Temporal ICA Models

1 Hidden Markov Independent Component Analysis

William D. Penny, Richard M. Everson and Stephen J. Roberts

1.1 Introduction

We propose a generative model for the analysis of non-stationary multivariate time series. The model uses a hidden Markov process to switch between independent component models where the components themselves are modelled as generalised autoregressive processes. The model is demonstrated on synthetic problems and EEG data. Much recent research in unsupervised learning [17,20] builds on the idea of using generative models [8] for modelling the probability distribution over a set of observations. These approaches suggest that powerful new data analysis tools may be derived by combining existing models using a probabilistic 'generative' framework. In this paper, we follow this approach and combine hidden Markov models (HMM), Independent Component Analysis (ICA) and generalised autoregressive models (GAR) into a single generative model for the analysis of non-stationary multivariate time series.

Our motivation for this work derives from our desire to analyse biomedical signals which are known to be highly non-stationary. Moreover, in signals such as the electroencephalogram (EEG), for example, we have a number of sensors (electrodes) which detect signals emanating from a number of cortical sources via an unknown mixing process. This naturally fits an ICA approach which is further enhanced by noting that the sources themselves are characterised by their dynamic content.

The overall generative model is shown in figure 1.1. We now look at each component of the model in turn. Section 1.2 describes the HMM, section 1.3 describes the ICA model and 1.3.1 and 1.3.2 describe the two different source models. Section 1.4 shows how HMMs and ICA can be integrated, section 1.5 discusses a number of practical issues and section 1.6 presents some empirical results.

1.2 Hidden Markov Models

The HMM model has K discrete hidden states, an initial state probability vector $\boldsymbol{\pi} = \pi_1, \pi_2, ..., \pi_K$, a state transition matrix $\boldsymbol{A}$ with entries a_{ij} and an observation density, $b_k(\boldsymbol{x}_t)$, for data point $\boldsymbol{x}_t$ and hidden state k. The

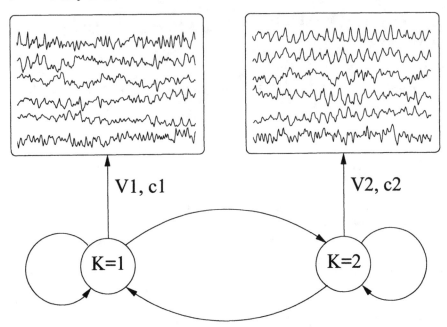

Fig. 1.1. Generative model. At time t the state of the HMM is $q_t = k$. Associated with each state is a mixing matrix $\boldsymbol{V}_k$ and a set of GAR coefficients, $\boldsymbol{c}_k^i$, one for each source i. The observations are generated as $\boldsymbol{x}_t = \boldsymbol{V}_k \boldsymbol{a}_t$ and the ith source is generated as $a_t^i = \boldsymbol{c}_k^i \tilde{\boldsymbol{a}}_t + e_t$ where $\tilde{\boldsymbol{a}}_t$ is a vector of previous source values and e_t is non-Gaussian noise.

parameters of the observation model (which in this paper, we consider to be an ICA model) are concatenated into the parameter set $\boldsymbol{B}$. The parameters of the HMM, as a whole, are concatenated into the parameter set $\boldsymbol{\theta} = \{\boldsymbol{A}, \boldsymbol{B}, \boldsymbol{\pi}\}$. The probability of being in state k at time t given a particular observation sequence $\boldsymbol{X}_1^T$ is $\gamma_k[t]$. Given $\boldsymbol{\theta}$, $\gamma_k[t]$ can be calculated from the forward-backward algorithm [16].

The HMM can be trained using an Expectation-Maximisation (EM) algorithm [7] as follows. In an HMM parameterised by some vector $\hat{\boldsymbol{\theta}}$ the observed variable $\boldsymbol{x}_t$ depends *only* on the current state q_t (and not previous states) and the current state depends *only* on the previous state (and not states before that). This can be seen from the generative model in figure. 1. This allows the joint likelihood of an observation sequence $\boldsymbol{x}_1^T = [\boldsymbol{x}_1, \boldsymbol{x}_t, ..., \boldsymbol{x}_T]$ and hidden variable (state) sequence $\boldsymbol{q}_1^T = [q_1, q_2, ..., q_T]$ to be written as

$$p_{\hat{\boldsymbol{\theta}}}(\boldsymbol{q}_1^T, \boldsymbol{x}_1^T) = p_{\hat{\boldsymbol{\theta}}}(q_1) \prod_{t=2}^{T} p_{\hat{\boldsymbol{\theta}}}(q_t \mid q_{t-1}) \prod_{t=1}^{T} p_{\hat{\boldsymbol{\theta}}}(\boldsymbol{x}_t \mid q_t) \tag{1.1}$$

The joint log likelihood may be written

$$\log p_{\hat{\boldsymbol{\theta}}}(\boldsymbol{q}_1^T, \boldsymbol{x}_1^T) = \log p_{\hat{\boldsymbol{\theta}}}(q_1) + \sum_{t=2}^{T} \log p_{\hat{\boldsymbol{\theta}}}(q_t \mid q_{t-1}) + \sum_{t=1}^{T} \log p_{\hat{\boldsymbol{\theta}}}(\boldsymbol{x}_t \mid q_t)$$

(1.2)

The EM algorithm requires us to maximise an *auxiliary function*, $Q(\boldsymbol{\theta}, \hat{\boldsymbol{\theta}})$, the expectation of the joint log likelihood where the expectation is taken relative to the old distribution of hidden (state) variables, $p_{\boldsymbol{\theta}}(\boldsymbol{q}_1^T \mid \boldsymbol{x}_1^T)$. [1] Hence,

$$Q(\boldsymbol{\theta}, \hat{\boldsymbol{\theta}}) = Q(\boldsymbol{\pi}, \hat{\boldsymbol{\pi}}) + Q(\boldsymbol{A}, \hat{\boldsymbol{A}}) + Q(\boldsymbol{B}, \hat{\boldsymbol{B}})$$

(1.3)

where

$$Q(\boldsymbol{\pi}, \hat{\boldsymbol{\pi}}) = \sum_{\boldsymbol{q}_1^T} p_{\boldsymbol{\theta}}(\boldsymbol{q}_1^T \mid \boldsymbol{x}_1^T) \log p_{\hat{\boldsymbol{\theta}}}(q_1)$$

(1.4)

$$Q(\boldsymbol{A}, \hat{\boldsymbol{A}}) = \sum_{\boldsymbol{q}_1^T} p_{\boldsymbol{\theta}}(\boldsymbol{q}_1^T \mid \boldsymbol{x}_1^T) \sum_{t=2}^{T} \log p_{\hat{\boldsymbol{\theta}}}(q_t \mid q_{t-1})$$

(1.5)

$$Q(\boldsymbol{B}, \hat{\boldsymbol{B}}) = \sum_{\boldsymbol{q}_1^T} p_{\boldsymbol{\theta}}(\boldsymbol{q}_1^T \mid \boldsymbol{x}_1^T) \sum_{t=1}^{T} \log p_{\hat{\boldsymbol{\theta}}}(\boldsymbol{x}_t \mid q_t)$$

(1.6)

and $\sum_{\boldsymbol{q}_1^T}$ denotes a sum over all possible hidden state sequences. The three terms can be maximised separately giving rise to parameter update equations for each part of the model. Maximising $Q(\boldsymbol{\pi}, \hat{\boldsymbol{\pi}})$ and $Q(\boldsymbol{A}, \hat{\boldsymbol{A}})$ leads to the usual update equations for the initial state probabilities and the state transition matrix [16]. The term for the observation model parameters can be re-arranged as

$$Q(\boldsymbol{B}, \hat{\boldsymbol{B}}) = \sum_{k} \sum_{t} \gamma_k[t] \log p_{\hat{\boldsymbol{\theta}}}(\boldsymbol{x}_t \mid q_t)$$

(1.7)

In principle, the observation model can be any probabilistic model. In speech recognition, for example, a Gaussian observation model or Gaussian Mixture model is often used and in previous work we have investigated the use of autoregressive models [14]. In this paper, we consider ICA models.

After the HMM has been trained, it can be applied to novel data sets and the most likely sequence of hidden states calculated using the Viterbi algorithm [16].

---

[1] Maximising the auxiliary function can be shown, via Baye's rule and Jensen's inequality, to maximise the likelihood of the data, $p_{\hat{\boldsymbol{\theta}}}(\boldsymbol{x}_t)$ [7].

1.3   Independent Component Analysis

In Independent Component Analysis (ICA) the observed variable is modelled as

$$x_t = V a_t \tag{1.8}$$

where the underlying factors or sources, a_t, are statistically independent and V is known as the *mixing matrix*. There are N observations and M sources. In this paper we consider square mixing only, i.e. $M = N$. ICA may also be viewed as a linear transformation to a new set of variables

$$a_t = W x_t \tag{1.9}$$

whose components, $a_i[t]$, are statistically independent. The matrix W is known as the *unmixing matrix* and is the inverse of V.

As noted by MacKay [11], it is possible to place ICA in a maximum likelihood framework. The likelihood of data points can be calculated from the likelihood of the source densities by noting that multivariate probability distributions are transformed as follows

$$p(x_t) = \frac{p(a_t)}{|J|} \tag{1.10}$$

where $||$ denotes the absolute value and J is the Jacobian of the transformation (the Jacobian is the determinant of the matrix of partial derivatives) and $a_t = W x_t$ [32]. Given that, in ICA, the transformation is linear and we are assuming independent sources, we have $|J| = 1/|det(W)|$ and $p(a_t) = \prod_i a_i[t]$. This allows us to write the log likelihood as

$$\log p(x_t) = \log |det(W)| + \sum_{i=1}^{M} \log p(a_i[t]) \tag{1.11}$$

where W is the unmixing matrix and $p(a_i[t])$ is the probability density of the ith source component and $a_i[t] = \sum_j W_{ij} x_j[t]$. We now describe two source density models.

1.3.1   Generalised Exponential Sources

In Bell and Sejnowski's original ICA formulation [4] the source densities were (implicitly) assumed to be inverse-cosh densities (see [11] for a discussion). More recent work (see, for example, [9]) however, shows that the inverse-cosh form cannot adequately model sub-Gaussian densities.[2] A more general parametric form which can model super-Gaussian, Gaussian and sub-Gaussian

---

[2] A sub-Gaussian density has negative kurtosis, a Gaussian density zero kurtosis, and a super-Gaussian density positive kurtosis. Sub-Gaussian densities have lighter tails and super-Gaussian densities have heavier tails than Gaussians.

forms is the 'Exponential Power Distribution' or 'Generalised Exponential' (GE) density

$$G(a; R, \beta) = \frac{R\beta^{1/R}}{2\Gamma(1/R)} \exp(-\beta|a|^R) \tag{1.12}$$

where $\Gamma()$ is the gamma function [15]. This density has zero mean, a kurtosis determined by the parameter R and a variance which is then determined by $1/\beta$; see the Appendix for further details. This source model has been used in an ICA model developed by Everson and Roberts [9]. The unmixing matrix, W, can be estimated using the usual covariant-based algorithms, or second-order methods such as BFGS (Broyden, Fletcher, Goldfarb, Shanno), and the source density parameters, $\{R, \beta\}$, can be estimated using an embedded line search [9].

With an explicit model for the source density, such as a generalised exponential, it is convenient for interpretation purposes for the unmixing matrix to be row-normalised. This allows the *directions* of the unmixing to be solely contained in W and the *magnitudes* of the sources to be solely contained in the source density parameters, eg. R and β. For all of the simulations in this paper we use row normalisation.

1.3.2   Generalised Autoregressive Sources

Pearlmutter and Parra [13] have proposed a 'contextual-ICA' algorithm where the sources are conditioned on previous source values. The observations are then generated from an instantaneous mixing of the sources. Their work focuses on using generalised autoregressive (GAR) models for modelling each source. The term 'generalised' is used because the AR models incorporate additive noise which is non-Gaussian. Specifically, Pearlmutter and Parra use an inverse-cosh noise distribution.

Pearlmutter and Parra have shown that contextual-ICA can separate sources which cannot be separated by standard (non-contextual) ICA algorithms. This is because the standard methods utilise only information from the cumulative histograms; temporal information is discarded.

These methods are particularly suited to EEG, for example, as sources are identified both by their spatial location and by their temporal content e.g. frequency characteristics.

In this paper we will use contextual information by modelling the sources as GAR models with p filter taps and additive noise drawn from a generalised exponential distribution; for $p = 0$ these models therefore reduce to the GE sources described in the previous section. The density model is

$$p(a_i[t]) = G(e_i[t], R_i, \beta_i) \tag{1.13}$$

where $e_i[t] = a_i[t] - \hat{a}_i[t]$ is the GAR prediction error and $\hat{a}_i[t]$ is the GAR prediction

$$\hat{a}_i[t] = -\sum_{d=1}^{p} c_i[d]a_i[t-d] \tag{1.14}$$

where $c_i[d]$ are the GAR coefficients for the ith source which can collectively be written as a vector $\boldsymbol{c}_i$.

1.4   Hidden Markov ICA

A Hidden Markov ICA model has an unmixing matrix, $\boldsymbol{W}_k$, and source density parameter vectors, $\boldsymbol{R}_k, \boldsymbol{\beta}_k$ for each state k. These parameters can be optimised by minimising a cost function defined by substituting the ICA log-likelihood (equation 1.11) into the HMM auxiliary function (equation 1.7)

$$Q = \sum_k Q_k \tag{1.15}$$

where

$$Q_k = \log |det(\boldsymbol{W}_k)| + \frac{1}{\gamma_k} \sum_t \gamma_k[t] \sum_i \log p(a_i[t]) \tag{1.16}$$

where we have divided through by

$$\gamma_k = \sum_t \gamma_k[t] \tag{1.17}$$

Gradient-based learning rules for updating the parameters of the HMICA model can be derived by evaluating the appropriate derivatives of Q_k e.g. $\frac{\partial Q_k}{\partial \boldsymbol{W}_k}$, $\frac{\partial Q_k}{\partial R_{k,i}}$, $\frac{\partial Q_k}{\partial \beta_{k,i}}$ and $\frac{\partial Q_k}{\partial \boldsymbol{c}_{k,i}}$. The remainder of this section shows how these derivatives can be computed. The unmixing matrix can be learnt by following the gradient

$$\frac{\partial Q_k}{\partial \boldsymbol{W}_k} = \boldsymbol{W}_k^{-T} + \frac{1}{\gamma_k} \sum_t \gamma_k[t] z_t x_t^T \tag{1.18}$$

where $z_i(t) = \partial \log p(a_i[t])/\partial a_i[t]$. It is, however, more efficient to use a ' covariant' [11] algorithm which follows the ' natural gradient' [1] which is obtained by post-multiplying the gradient by $\boldsymbol{W}_k^T \boldsymbol{W}_k$. This gives

$$\left[\frac{\partial Q_k}{\partial \boldsymbol{W}_k}\right]_{NAT} = \boldsymbol{W}_k + \frac{1}{\gamma_k} \sum_t \gamma_k[t] z_t y_t^T \tag{1.19}$$

where

$$\boldsymbol{y}_t = \boldsymbol{W}_k^T \boldsymbol{a}_t \tag{1.20}$$

The above gradients are identical to those used in standard ICA except that, in the ICA update for state k, each data point is weighted by an amount proportional to the 'responsibility' state k has for that data point, $\gamma_k[t]/\gamma_k$.

In what follows, there is an unmixing matrix for each state and there are source model parameters for each source and for each state. But, to keep the notation simple, we drop any references to the state k (e.g. we write W_{ij} not $W_{k,ij}$, R_i not $R_{k,i}$ and β_i not $\beta_{k,i}$). We now derive HMICA learning rules for the cases of both GE sources and GAR sources.

1.4.1 Generalised Exponential Sources

The natural gradient is given by substituting the derivative of the GE density into the general weight update equation derived above. We therefore substitute $z_i(t)$ from equation 1.35 into equation 1.19. This gives

$$\left[\frac{\partial Q_k}{\partial W_{ij}}\right]_{NAT} = W_{ij} - \frac{1}{\gamma_k}\sum_t \gamma_k[t]sign(a_i[t])R_i\beta_i|a_i[t]|^{R_i-1}y_j[t] \qquad (1.21)$$

where

$$y_j[t] = \sum_i W_{ij}a_i[t] \qquad (1.22)$$

To update the parameters of the GE sources the appropriate likelihood function can found by substituting $\log p(a_i[t])$ from equation 1.34, for each source in turn, into equation 1.16. This gives

$$Q_k = \log|det(W_k)| + \frac{1}{\gamma_k}\sum_t \gamma_k[t]\sum_i \log\left[\frac{R_i\beta_i^{1/R_i}}{2\Gamma(1/R_i)}\right] - \beta_i|a_i(t)|^R \qquad (1.23)$$

The source density parameters for each state can now be estimated component-wise by evaluating the derivatives $\frac{\partial Q_k}{\partial \beta_i}$ and $\frac{\partial Q_k}{\partial R_i}$, where i refers to the ith source, and setting them to zero. This gives

$$\frac{1}{\beta_i} = \frac{R_i}{\gamma_k}\sum_t^T \gamma_k[t]|a_i[t]|^{R_i} \qquad (1.24)$$

and R_i can be found by solving the one-dimensional problem

$$\frac{\partial Q_k}{\partial R_i} = \frac{1}{R_i} + \frac{1}{R_i^2}\log\beta_i + \frac{1}{R_i^2}\varphi(1/R_i) - \frac{\beta_i}{\gamma_k}\sum_{t=1}^T \gamma_k[t]|a_i[t]|^{R_i}\log|a_i[t]| = 0$$

$$(1.25)$$

where $\varphi(x) = \Gamma'(x)/\Gamma(x)$ is the digamma function [15]. Equations 1.24 and 1.25 are generalised versions of the equations in Appendix A of [9] and are applied iteratively until a consistent solution is reached.

1.4.2   Generalised Autoregressive Sources

For GAR sources the analysis is identical to that described for GE sources except that, in equations 1.21 to 1.25 (but not equation 1.22), $a_i[t]$ is replaced by $e_i[t]$ where $e_i[t] = a_i[t] - \hat{a}_i[t]$ and $\hat{a}_i[t]$ is the GAR prediction. The GAR coefficients, c_i, can be found separately for each source by minimising

$$E_{k,i} = \frac{1}{\gamma_k} \sum_t^T \gamma_k[t]|a_i[t] - \hat{a}_i[t]|^{R_i} \tag{1.26}$$

which can be achieved by descending the gradient $\frac{\partial E_{k,i}}{\partial c_i}$. For the dth element of c_i this gives

$$\frac{\partial E_k}{\partial c_{d,i}} = \frac{1}{\gamma_k} \sum_t \gamma_k[t]a_i[t-d]R_i sign(e_i[t])|e_i[t]|^{R_i-1} \tag{1.27}$$

We use this gradient in a BFGS optimizer [15].

1.5   Practical Issues

1.5.1   Initialisation

As learning is based on a maximum likelihood framework, and there may be a number of local maxima in the likelihood landscape, our final solution may be sensitive to the initial parameter settings.

Domain knowledge can be used to initialise some of the parameters. For example, it might be known that some states persist for a stereotypical time period. This information can be used to initialise the state transition matrices. It may also be known that the sources we are interested in have a particular spectral content or are in a particular location. This information can be used to initialise the GAR source models or ICA unmixing matrices.

Alternatively, we could use methods employed in previous research [14] where a first pass through the data and a subsequent cluster analysis identifies good initial conditions for the observation models.

We may also have domain knowledge which requires that some parameters be tied or fixed to certain values. For example, it may be known that the sources do not change but the mixing matrices do, or vice-versa.

1.5.2   Learning

The pseudo-code in figure 1.2 shows the three layers of inference in the HMICA model with GAR sources. The three nested loops (HMM optimisation, ICA optimisation and GAR optimisation) require convergence criteria

Initialise HMM transition matrix, $\boldsymbol{A}$, and initial probability vector, $\boldsymbol{\pi}$.

Initialise ICA mixing matrices, $\boldsymbol{W}_k$, and source parameters c_i, R_i and β_i.

While HMM updates not converged,

    Run forward-backward algorithm to calculate $\gamma_k[t]$.

    For k = 1 to K,

        While ICA updates not converged,

            Get source values, predicted source values and errors:

$$a_t = \boldsymbol{W}\boldsymbol{x}_t, \hat{a}_i[t] = -\sum_{d=1}^{p} c_i[d]a_i[t-d], e_i[t] = a_i[t] - \hat{a}_i[t]$$

            Update unmixing matrix:

$$\Delta\boldsymbol{W}_{ij} \propto -\frac{1}{\gamma_k}\sum_t \gamma_k[t]sign(e_i[t])R_i\beta_i|e_i[t]|^{R_i-1}y_j[t]$$

            If ICA iteration number is divisible by S,

                Update GAR coefficients:

$$\Delta c_{i,d} \propto \frac{1}{\gamma_k}\sum_t \gamma_k[t]a_i[t-d]R_i sign(e_i[t])|e_i[t]|^{R_i-1}$$

                Update GAR noise model:

$$\frac{1}{\beta_i} = \frac{R_i}{\gamma_k}\sum_t^T \gamma_k[t]|e_i[t]|^{R_i}. \text{ Also update } R_i.$$

            end

        end

    end

    Update HMM state transition matrix and initial probability vector.

end

Fig. 1.2. Pseudo-code for learning in a hidden Markov ICA model with GAR sources.

which are defined as follows. If $LL(new)$, $LL(old)$ and $LL(init)$ are the old, new and initial log-likelihood values then convergence is defined by

$$\frac{LL(new) - LL(init)}{LL(old) - LL(init)} < 1 + \delta \tag{1.28}$$

For the HMM loop we use $\delta = 10^{-4}$ and for the ICA and GAR loops we use $\delta = 10^{-2}$. A further issue is the choice of S, the number of updates of the ICA unmixing matrix before the source model is re-estimated. In this paper we set $S = 10$. Optimisation does not, however, seem to be particularly sensitive to the choice of these parameters.

1.5.3   Model Order Selection

We have to decide on the number of HMM states, K, the number of source variables, M, and the number of GAR parameters in each source model, p. In general, we rely on domain knowledge, although generic methods do exist, for example, for GAR model order selection [6] and HMM model order selection [18]. For the ICA model, as we consider only the case of square mixing, model order selection is not an issue, although we discuss this further in section 1.7.

1.6   Results

In section 1.6.1 we compare the GE and GAR source models on a synthetic data set. In 1.6.2 and 1.6.3 we demonstrate that the HMICA model can partition multiple time series on the basis of having different underlying source models or mixing matrices or both. In 1.6.4 we apply the HMICA model to EEG data.

1.6.1   Multiple Sinewave Sources

In this section we compare the GE and GAR source models on a synthetic data set. The data consist of two sources, one composed of three sine waves of unit amplitude (ranging from +1 to -1) at frequencies of 5, 15 and 25 Hz plus additive Laplacian noise, and the other composed of two sine waves of unit amplitude at frequencies of 10 and 20 Hz plus additive Gaussian noise. [3] The two sources were then mixed using

$$V = \begin{pmatrix} 0.8321 \; 0.5547 \\ 0.4472 \; 0.8944 \end{pmatrix} \tag{1.29}$$

---

[3] If the sampling rate is set to be a multiple of any of the frequencies in the signal this can result in pathological sampling; care was therefore taken to avoid this by choosing the sampling rate to be $60 + \pi$Hz, an irrational number.

and $N = 200$ samples were generated in this way. The quality of the unmixing was measured by the unmixing error

$$E = \min_{P} \frac{\|WVP - diag(WVP)\|}{\|WV\|} \tag{1.30}$$

where W is our estimated unmixing matrix and P is a permutation matrix [9]. The variance of the source noise was then set so as to achieve a certain Signal-to-Noise Ratio (SNR), which was the same for both sources. Table 1.1 shows the unmixing errors for the two methods over a range of SNRs. The results show that, as the source SNR increases, the GAR source model

Table 1.1. Comparison of GE and GAR source models. Unmixing errors, E, for Generalised Exponential (GE) and Generalised Autoregressive (GAR) source models on multiple sine wave sources. Each entry is the mean value from 20 runs with $\pm\sigma$ indicating the standard deviation. At high source SNR, the GAR source model provides better unmixing.

SNR	E_{GE}	E_{GAR}
0.01	0.09 ± 0.05	0.13 ± 0.07
0.10	0.08 ± 0.05	0.10 ± 0.07
1.00	0.13 ± 0.09	0.06 ± 0.04
10.0	0.19 ± 0.29	0.02 ± 0.01

results in better unmixing; at $SNR = 10$ the unmixing is improved by a factor of 10. [4] The reason for this comparative improvement is two-fold; firstly, the GAR source model itself does a better job and, secondly, and more significantly, the GE model breaks down at high SNR. This breakdown is attributed to the fact that, at high SNR, a signal consisting of multiple sinusoids has a multimodal PDF. [5] This multimodality can also be seen empirically in figure 1.3. Therefore, the unimodal GE distribution is not capable of modelling high SNR sources. The GAR model, on the other hand, fits the dynamics of the source signals leaving a unimodal residual which is adequately modelled by its GE noise model.

---

[4] At low SNR, we effectively have a Laplacian source and a Gaussian source which can be separated by ICA with a GE source model [9]; if *both* sources were Gaussian, however, then this would not be the case.

[5] The PDF of a single sinusoid $R\sin wt$ has modes at $\pm R$ [32], and the addition of further sinusoids of different amplitude and frequency can result in further modes. This PDF is then convolved with the noise PDF to produce a somewhat smeared density where the amount of smearing depends on the noise variance.

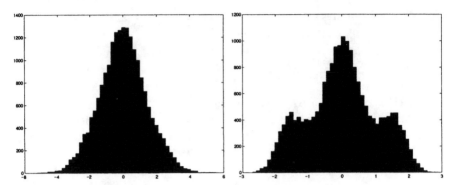

Fig. 1.3. Multimodality of multiple sine wave sources. Empirical PDF (N=20,000) of source 2 at SNR=1 (*left plot*) and SNR=10 (*right plot*). At lower SNRs the noise acts to smear out the peaks resulting in a unimodal PDF.

Knuth's empirical work [10] suggests that many EEG sources have multimodal PDFs. For such cases, we would therefore expect GAR source models to provide better unmixing than GE source models. Ultimately, as the GE source model is a special case of GAR with $p = 0$, the choice of which is more appropriate is a model order selection problem.

1.6.2   Same Sources, Different Mixing

We generated a two-dimensional time series containing two epochs each with $N = 200$ samples. We first generated two GE sources, the first being Gaussian ($R = 2$) with variance $1/\beta = 2$, and the second being Laplacian ($R = 1$) with width parameter $1/\beta = 1$. These source parameters were the same for both epochs. Observations for the first epoch of the time series were then generated by mixing the sources with the matrix

$$V_1 = \begin{pmatrix} 0.8321\ 0.5547 \\ 0.4472\ 0.8944 \end{pmatrix} \tag{1.31}$$

In the second epoch, the matrix was changed to

$$V_2 = \begin{pmatrix} +0.8944\ -0.4472 \\ -0.5547\ +0.8231 \end{pmatrix} \tag{1.32}$$

The resulting data set is plotted in figure 1.4. We then trained two ICA models on each separate data partition to serve as a benchmark against which the HMICA solution could be compared.[6]

---

[6] If we were to compare the HMICA solution against the known true models then any difference could be due to (i) incorrect partitioning in the HMICA model, (ii) having a small, finite data sample or (iii) some artifact of the covariant ICA algorithm. By comparing against the covariant ICA solutions we are therefore sure to be focusing only on (i).

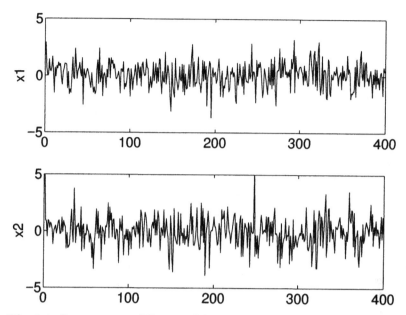

Fig. 1.4. Same sources, different mixing. The mixing matrices underlying the two time series are switched at t=200.

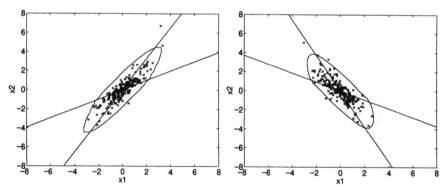

Fig. 1.5. Same sources, different mixing. ICA models for state 1 (*left plot*) and state 2 (*right plot*). The straight lines show the directions of the independent components and the ellipse shows a single probability contour plotted at 0.01 of the probability density at the origin. These probability densities were calculated using equation 1.10. The solid lines are from individual ICA models trained on the correct partitioning of the data and the dotted lines are estimates from the HMICA model (the dotted lines are barely visible as the estimates overlap almost exactly with the true values).

A two-state HMICA model was then trained on the data. The model was initialised by setting $R = 2$ and $\beta = 1$ for both sources and both states. Each unmixing matrix was initially set to the identity matrix with uniform noise in the range $[-0.05, 0.05]$ added on to each matrix element. The diagonal elements of the state transition matrix were set so that the average state duration density was D_s samples. Five iterations of the EM algorithm (using equations 1.21, 1.24 and 1.25 to update the ICA unmixing matrices and source density parameters) were then sufficient for the HMICA solution to converge. The estimated ICA models are compared to the benchmark ICA models in figure 1.5. The almost complete overlap of true and estimated probability contours shows that R and β are estimated very accurately. This accuracy is due to the correct partitioning of the data into the two separate epochs; application of the Viterbi algorithm led to 100% of samples being correctly classified. Experimentation with initialisation of the state transition matrix showed that convergence was typically achieved in five to ten EM iterations with D_s ranging from 10 to 400 samples. Convergence was even possible with D_s as low as three, after 10 to 20 EM iterations.

1.6.3   Same Mixing, Different Sources

We generated a two-dimensional time series containing two epochs each with $N = 200$ samples. Data for the first epoch was identical to the first epoch in section 1.6.2. Data for the second epoch used the same mixing matrix but the source densities were chosen to be uniform.

The resulting data set is plotted in Figure 1.6. We then trained two ICA models on each separate data partition to serve as a benchmark against which the HMICA solution could be compared. A two-state HMICA model was then trained on the data after initialising using the same method as in the previous example. Seven iterations of the EM algorithm were then sufficient for the HMICA solution to converge. The estimated ICA models are compared to the benchmark ICA models in figure 1.7. Application of the Viterbi algorithm led to 100% of samples being correctly classified.

1.6.4   EEG Data

We now apply the HMICA model to EEG data recorded whilst a subject performed voluntary hand movements in response to cues. Figure 1.8 shows two channels of EEG and the timing of the movement cue. This subject possessed a particularly strong mu-rhythm (an 8-12Hz rhythm in sensorimotor cortex) which disappeared during hand-movement (after the cue).

We trained two HMICA models on this data set; one with GAR sources, one with GE. Each converged within ten EM iterations. As can be seen from figure 1.8, the GAR sources provide a reasonable partitioning of the data, whereas use of the GE source model provides a less clear partitioning. This

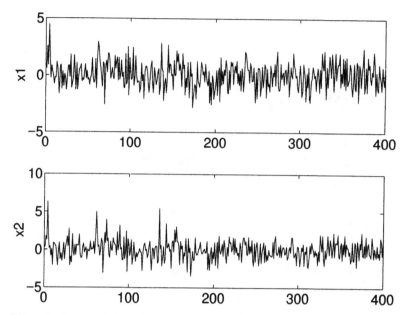

Fig. 1.6. Same mixing, different sources. The source models underlying the two time series are switched at t=200

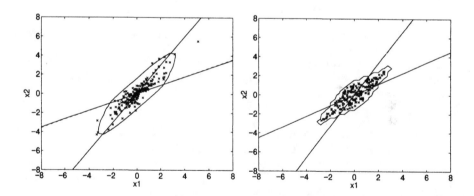

Fig. 1.7. Same mixing, different sources. ICA models for state 1 (*left plot*) and state 2 (*right plot*). The straight lines show the directions of the independent components and the ellipse shows a single probability contour plotted at 0.01 of the probability density at the origin. The solid lines are from individual ICA models trained on the correct partitioning of the data and the dotted lines are estimates from the HMICA model (the dotted lines are barely visible as the estimates overlap almost exactly with the true values).

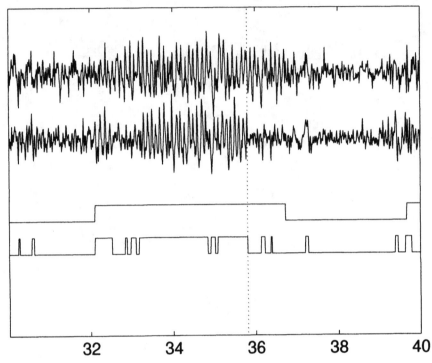

Fig. 1.8. EEG data recorded before and after motor task. The top two traces show EEG recordings from electrodes at positions C3 and C4 (10-20 System). The next two traces show partitionings from an HMICA model; the top one using GAR sources and the bottom one using GE sources. The dashed vertical line indicates a cue telling the subject when to begin hand movement.

difference can be explained by the high SNR during state 2 (see section 1.6.1); for the GAR sources in state 2, the SNRs are 3.4 and 5.5.

We also applied the HMICA model to EEG data recorded whilst a subject performed a motor imagery task; the subject was instructed to imagine opening and closing his hand in response to an external cue. Figure 1.9 shows the EEG data and the resulting partitioning from two HMICA models. This again demonstrates that the HMICA model can partition time series into meaningful stationary regimes and that GAR source models may often be more appropriate than GE source models.

We note that it is essential to have a non-stationary model in order for these periods of strong (high SNR) rhythmic source activity to be identified. Applying stationary ICA to the time series without the partitioning provided by the HMM misses this activity completely. GAR source models and HMM switching are therefore complementary.

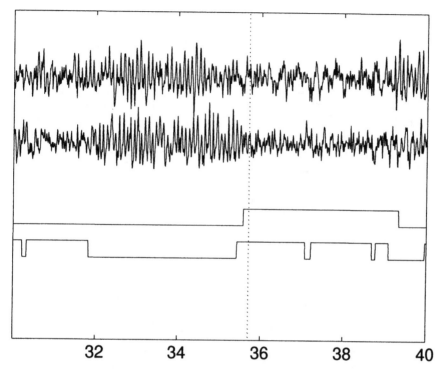

Fig. 1.9. EEG data recorded before and after motor imagery task. The top two traces show EEG recordings from electrodes at positions C3 and C4 (10-20 System). The next two traces show partitionings from an HMICA model; the top one using GAR sources and the bottom one using GE sources. The dashed vertical line indicates the cue indicating when the task should be performed.

1.7 Conclusion

We have proposed a generative model for the analysis of non-stationary multivariate time series. The model combines three existing statistical models, HMMs, ICA and GARs, into a single overall model where the components learn together. The model can (i) learn the dynamics of the underlying sources, (ii) learn the parameters of the mixing process and (iii) detect discrete changes in either the source dynamics or the mixing or both. This last property enables the model to partition a multivariate time series into different regimes and so may be used for classification purposes. Moreover, each regime is characterised by a known mixing of stationary sources. The model is therefore readily interpretable.

As compared to previous work in ICA, the main novelty of our approach is in the use of an HMM to switch between ICA models. Like other modellers [13], we have also used GAR source models, although ours are slightly different in the choice of noise model; instead of an inverse-cosh we use the more flexible generalised exponential [9].

Whilst our model goes some way to bridging the gap between statistical models and the known properties of biomedical data, such as EEG, there remain many areas for refinement. The most unrealistic assumption is that of square mixing; for data such as EEG, it is likely that the number of sources is less than the number of electrodes. A non-square ICA algorithm is therefore required, such as those developed by Attias[2,3]. These algorithms also have the advantage of being able to cope with observation noise (actually, the observation noise is essential in developing a probabilistic model; if you have fewer sources than observations the 'extra' observations can be explained as noise).

The second main area for improvement is the use of automatic model order selection methods; both for the order of the source model and to choose the optimal number of sources. Again, Attias' algorithms can do this. Finally, following Knuth [10], we could use our knowledge of the physics of the mixing process to choose suitable priors for the mixing matrices.

1.8   Acknowledgements

William Penny is supported by the UK Engineering and Physical Sciences Research Council grant number GR/M 41339. The empirical work in this paper was based around hidden Markov model software routines made available by Zoubin Ghahramani (http://gatsby.ucl.ac.uk/~zoubin).

1.9   Appendix

The 'exponential power' or 'generalised exponential' probability density is defined as

$$p(a) = G(a; R, \beta) = \frac{R\beta^{1/R}}{2\Gamma(1/R)} \exp(-\beta|a|^R) \tag{1.33}$$

where $\Gamma()$ is the usual gamma function [15], the mean of the distribution is zero, the width of the distribution is determined by $1/\beta$ and the weight of its tails is set by R. This gives rise to a Gaussian distribution for $R = 2$, a Laplacian for $R = 1$ and a uniform distribution in the limit $R \to \infty$. The log-likelihood is given by

$$\log p(a) = \log\left[\frac{R\beta^{1/R}}{2\Gamma(1/R)}\right] - \beta|a|^R \tag{1.34}$$

The partial derivative, $z(a) = \frac{\partial \log p(a)}{\partial a}$, is given by

$$z(a) = -sign(a)R\beta|a|^{R-1} \tag{1.35}$$

We also note that the density is equivalently parameterised by a variable w where $w = \beta^{-1/R}$ giving

$$p(a) = \frac{R}{2w\Gamma(1/R)} \exp(-|a/w|^R) \qquad (1.36)$$

and that the density is unimodal and symmetric with variance given by

$$V = w^2 \frac{\Gamma(3/R)}{\Gamma(1/R)} \qquad (1.37)$$

which for $R = 2$ gives $V = 0.5w^2$. The kurtosis is given by [5]

$$K = \frac{\Gamma(5/R)\Gamma(1/R)}{\Gamma(3/R)^2} - 3 \qquad (1.38)$$

Samples may be generated from the density using a rejection method [19].

References

1. S. Amari, A. Cichocki, and H. H. Yang. A new learning algorithm for blind signal separation. In D. S. Touretzky, M. C. Mozer, and M. E. Hasselmo, editors, *Advances in Neural Information Processing Systems 8*. MIT Press, Cambridge, MA, 1996.
2. H. Attias. Independent factor analysis. *Neural Computation*, **11**(4):803–851, 1999.
3. H. Attias. Inferring parameters and structure of latent variable models by variational Bayes. In *Proceedings of the Fifteenth Conference on Uncertainty in Artificial Intelligence*, 1999. http://www.gatsby.ucl.ac.uk/~hagai/papers.html
4. A. J. Bell and T. J. Sejnowski. An information-maximization approach to blind separation and blind deconvolution. *Neural Computation*, **7**(6):1129–1159, 1995.
5. P. Bratley, B. L. Fox, and E. L. Schrage. *A guide to simulation*. Springer-Verlag, 1983.
6. C. Chatfield. *The analysis of time series: An introduction*. Chapman and Hall, 1996.
7. A. P. Dempster, N. M. Laird, and D. B. Rubin. Maximum likelihood from incomplete data via the EM algorithm. *Journal of the Royal Statistical Society-B*, **39**:1–38, 1977.
8. B. S. Everitt. *An introduction to latent variable models*. Chapman and Hall, London and New York, 1984.
9. R. Everson and S. J. Roberts. Independent component analysis: A flexible nonlinearity and decorrelating manifold approach. *Neural Computation*, **11**(8), 1999.
10. K. H. Knuth. Difficulties applying recent blind source separation techniques to EEG and MEG. In J. T. Rychert and C. R. Smith, editors, *Maximum Entropy and Bayesian Methods*, 209–222. Dordrecht, 1998.

11. D. J. C. MacKay. Maximum likelihood and covariant algorithms for independent component analysis. Technical report, Cavendish Laboratory, University of Cambridge, 1996.
12. A. Papoulis. *Probability, Random Variables, and Stochastic Processes.* McGraw-Hill, 1991.
13. B. A. Pearlmutter and L. C. Parra. Maximum likelihood blind source separation: A context-sensitive generalization of ICA. In *Advances in Neural Information Processing Systems 9*, 613–619. MIT Press, Cambridge, MA, 1997.
14. W. D. Penny and S. J. Roberts. Dynamic models for nonstationary signal segmentation. *Computers and Biomedical Research*, 1999. To appear.
15. W. H. Press, S. A. Teukolsky, W. T. Vetterling, and B. V. P. Flannery. *Numerical Recipes in C, second edition.* Cambridge, 1992.
16. L. R. Rabiner. A tutorial on hidden Markov models and selected applications in speech recognition. *Proceedings of the IEEE*, 77(2):257–286, 1989.
17. S. Roweis and Z. Ghahramani. A unifying review of linear Gaussian models. *Neural Computation*, 11(2):305–346, 1999.
18. T. Ryden. Estimating the order of hidden Markov models. *Statistics*, 26(4):345–354, 1995.
19. P. R. Tadikamalla. Random sampling from the exponential power distribution. *Journal of the American Statistical Association*, 75:683–686, 1980.
20. M. E. Tipping and C. M. Bishop. Mixtures of probabilistic principal component analyzers. *Neural Computation*, 11(2):443–482, 1999.

2 Particle Filters for Non-Stationary ICA

Richard M. Everson and Stephen J. Roberts

2.1 Introduction

Over the last decade in particular there has been much interest in Independent Component Analysis (ICA) methods for blind source separation (BSS) and deconvolution (see [15] for a review). One may think of the blind source separation as the problem of identifying speakers (sources) in a room given only recordings from a number of microphones, each of which records a linear mixture of the sources, whose statistical characteristics are unknown. The casting of this problem in a neuro-mimetic framework [3] has done much to to simplify and popularise the technique. More recently still the ICA solution has been shown to be the maximum-likelihood point of a latent-variable model [17,4,20]

Here we consider the blind source separation problem when the mixing of the sources is non-stationary. Pursuing the analogy of speakers in a room, we address the problem of identifying the speakers when they (or equivalently, the microphones) are moving. The problem is cast in terms of a hidden state (the mixing proportions of the sources) which we track using particle filter methods, which permit the tracking of arbitrary state densities.

We first briefly review classical independent component analysis. ICA with non-stationary mixing is described in terms of a hidden state model and methods for estimating the sources and the mixing are described. Particle filter techniques are introduced for the modelling of state densities. Finally, we address the non-stationary mixing problem when the sources are independent, but possess temporal correlations.

2.2 Stationary ICA

Classical ICA belongs to the class of latent variable models which model observed data as being generated by the linear mixing of latent sources. We assume that there are M sources whose probability density functions (pdf) are $p_m(s^m)$. Observations, $\mathbf{x}_t \in \mathbb{R}^N$, are produced by the instantaneous linear mixing of the sources by A:

$$\boldsymbol{x}_t = A\boldsymbol{s}_t \tag{2.1}$$

In order for the problem to be well posed, the mixing matrix, A, must have at least as many rows as columns ($N \geq M$), so that the dimension of each observation is at least as great as the number of sources. The aim of ICA methods is to recover the latent sources $\hat{s}_t$ by finding W, the (pseudo-) inverse of A:

$$\hat{s}_t = W x_t = W A s_t \tag{2.2}$$

The probability density of the observations is related [19] to the joint source density by:

$$p(x_t | A) = \frac{p(s_t)}{|\det A|} \tag{2.3}$$

Hence the log likelihood of the observation x_t is

$$\log l = -\log|\det A| + \log p(s_t) \tag{2.4}$$

and the normalised log likelihood of a set of observations $t = 1, ..., T$ is therefore

$$\log \mathcal{L} = -\log|\det A| + \frac{1}{T} \sum_{t=1}^{T} \log p(s_t) \tag{2.5}$$

To proceed further some assumptions or prior knowledge about the nature of $p(s_t)$ must be incorporated into the model. Physical considerations may provide strong constraints on the form of $p(s_t)$ (see, for example, Miskin and MacKay, this volume). The assumption that the sources are linearly decorrelated leads to an $M(M-1)$-dimensional manifold of separating matrices [6]. The further constraint that the sources are normally distributed with zero mean and unit variance leads to the principal component analysis, PCA [12].

Independent component analysis assumes that the sources are independent so that the joint pdf of the sources factorises into the product of marginal densities:

$$p(s_t) = \prod_{m=1}^{M} p(s_t^m) \tag{2.6}$$

Using this factorisation, the normalised log likelihood of a set of T observations is therefore [4,17,20]:

$$\log \mathcal{L} = -\log|\det A| + \frac{1}{T} \sum_{t=1}^{T} \sum_{m=1}^{M} \log p_m(\hat{s}_t^m) \tag{2.7}$$

The optimum A may then be found by maximisation of $\log \mathcal{L}$ with respect to A, assuming some specific form for the marginal densities $p(s_t^m)$.

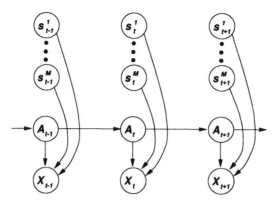

Fig. 2.1. The Graphical representation describing the non-stationary ICA model.

Sucessive gradient ascents on $\log l_t$ (equation 2.4) leads to the Bell and Sejnowksi stochastic learning rule for ICA [3], while batch learning is achieved by maximising $\log \mathcal{L}$. Learning rates may be considerably enhanced by modifying the learning rule to make it covariant [1,17]. A common choice is $p(s_t^m) \propto 1/\cosh(s_t^m)$, which leads to a tanh nonlinearity in the learning rule. Although the source model is apparently fixed, scaling of the mixing matrix tunes the model to particular sources [6] and, with a tanh nonlinearity, platykurtic (heavy-tailed) sources can be separated, although not leptokurtic ones. Cardoso [5] has elucidated the conditions under which the true mixing matrix is a stable fixed point of the learning rule.

Adoption of a more flexible model for the source densities permits the separation of a wider range of source densities. Attias [2] has used mixtures of Gaussians to model the sources, which permits multi-modal sources, and Lee *et al.* [16] switch between sub- and super-Gaussian source models. In the work presented here a generalised exponential model is used.

We emphasise that since the likelihood is unchanged if A is pre-multiplied by a diagonal matrix D or a scaling matrix P, the original scale of the sources cannot be recovered. The separating matrix W is therefore only the inverse of A up to a diagonal scaling and permutation, that is:

$$WA = PD \tag{2.8}$$

2.3   Non-Stationary Independent Component Analysis

Figure 2.1 depicts the graphical model describing the conditional independence relations of the non-stationary ICA model. In common with static independent component analysis, we adopt a generative model in which M independent sources are linearly mixed at each instant. Unlike static ICA, however, the mixing matrix A_t is allowed to vary with time. We also assume that the observation x_t is contaminated by normally distributed noise

$w_t \sim \mathcal{N}(0, R)$. Thus

$$x_t = A_t s_t + w_t \tag{2.9}$$

Since the observational noise is assumed to be Gaussian its density is

$$p(w_t) = \mathcal{G}(w_t, R) \tag{2.10}$$

where $\mathcal{G}(\cdot, \Sigma)$ denotes a Gaussian function with zero mean and covariance Σ. The pdf of observations $p(x_t \mid A_t)$ is given by

$$p(x_t \mid A_t) = \int p(x_t \mid A_t, \theta, s_t) p(s_t \mid \theta) \, ds_t \tag{2.11}$$

where θ are parameters pertaining to the particular source model. Since the *sources* are assumed stationary, the likelihood of x_t given a particular mixing matrix is

$$p(x_t \mid A_t) = \int p(x_t \mid A_t, s) p(s \mid \theta) \, ds \tag{2.12}$$

As for static ICA, the assumption of the independence of the sources permits the source density to be written in factorised form (equation 2.6) so that the likelihood becomes

$$p(x_t \mid A_t) = \int \mathcal{G}(x_t - A_t s, R) \prod_{m=1}^{M} p_m(s^m) \, ds \tag{2.13}$$

We emphasise that it is in equation 2.13 that the independence of the sources is modelled by writing the joint source density in factored form.

The dynamics of A_t are modeled by a first order Markov process, in which the elements of A_t diffuse from one observation time to the next. If we let $a_t = \text{vec}(A_t)$ be the $N \times M$-dimensional vector obtained by stacking the columns of A_t, then a_t evolves according to the state equation:

$$a_{t+1} = F a_t + v_t \tag{2.14}$$

where v_t is zero-mean Gaussian noise with covariance Q, and F is the state transition matrix; in the absence of *a priori* information we take F to be the identity matrix. The state equation 2.14 and the statistics of v_t together define the density $p(a_{t+1} \mid a_t)$.

A full specification of the hidden state must include the parameter set $\theta = \{\theta_m\}$, $m = 1...M$, which describes the independent source densities. We model the source densities with generalised exponentials, as described in section 2.3.1. Since the sources themselves are considered to be stationary, the parameters θ are taken to be static, but they must be learned as data are observed.

The problem is now to track A_t and to learn θ as new observations x_t become available. If X_t denotes the collection of observations $\{x_1, ..., x_t\}$, then the goal of filtering methods is to deduce the probability density function of the state $p(a_t | X_t)$. This pdf may be found recursively in two stages: prediction and correction. If $p(a_{t-1} | X_{t-1})$ is known, the state equation 2.14 and the Markov property that a_t depends only on a_{t+1} permits prediction of the state at time t:

$$p(a_t | X_{t-1}) = \int p(a_t | a_{t-1}) p(a_{t-1} | X_{t-1}) \, da_{t-1} \qquad (2.15)$$

The predictive density $p(a_t | X_{t-1})$ may be regarded as an estimate of a_t prior to the observation of x_t. As the datum x_t is observed, the prediction may be corrected via Bayes' rule

$$p(a_t | X_t) = Z^{-1} p(x_t | a_t) p(a_t | X_{t-1}) \qquad (2.16)$$

where the likelihood of the observation given the mixing matrix, $p(x_t | a_t)$, is defined by the observation equation 2.9. The normalisation constant Z is known as the innovations probability:

$$Z = p(x_t | X_{t-1}) = \int p(x_t | a_t) p(a_t | X_{t-1}) \, da_t \qquad (2.17)$$

The prediction 2.15 and correction/update 2.16 pair of equations may be used to step through the data online, alternately predicting the subsequent state and then correcting the estimate when a new datum is observed. The corrected state at time t is then the basis for the prediction of the state at $t + 1$.

2.3.1   Source Model

Before describing the application of particle filters to the prediction and correction equations we briefly describe the source model. In order to be able to separate light-tailed sources a more flexible source model than the traditional $1/\cosh$ density is needed. It is difficult to use a switching model [16] in this context. Attias [2] has used mixtures of Gaussians to model the sources, which permits multi-modal sources; we use generalised exponentials, which provide a good deal of flexibility and do not suffer from the combinatorial complexities associated with mixture models.

Each source density is modelled by

$$p(s^m | \theta_m) = z \exp - \left| \frac{s^m - \nu_m}{w_m} \right|^{r_m} \qquad (2.18)$$

where the normalising constant is

$$z = \frac{r_m}{2 w_m \Gamma(1/r_m)} \qquad (2.19)$$

The density depends upon parameters $\boldsymbol{\theta}_m = \{\mu_m, w_m, r_m\}$. The location of the distribution is set by μ_m, its width by w_m and the weight of its tails is determined by r_m. Clearly p is Gaussian when $r_m = 2$, Laplacian when $r_m = 1$, and the uniform distribution is approximated in the limit $r_m \to \infty$.

Generalised exponential source models in static ICA are able to separate mixtures of Laplacian distributed, Gaussian distributed and uniformly distributed sources and mixtures that methods using a static tanh nonlinearity are unable to separate [6].

2.4   Particle Filters

Practical implementation of the filtering equations 2.15 and 2.16 requires a representation of the predicted and corrected state densities: $p(\boldsymbol{a}_t | X_{t-1})$ and $p(\boldsymbol{a}_t | X_t)$. State densities in the Kalman filter [13,11], whose observation equation is linear, are Gaussian and thus easily represented. Nonlinear and non-Gaussian tracking problems must either approximate the state density by a tractable form (usually Gaussian) or use a sampling method. Particle filters, which date back to the Sampling Importance Resampling (SIR) filter of Gordon et al. [9], represent the state density $p(\boldsymbol{a}_t | X_{t-1})$ by a collection or swarm of "particles" each with a probability mass. Each particle's probability mass is modified using the state and observation equations after which a new independent sample is obtained from the posterior $p(\boldsymbol{a}_t | X_t)$ before proceeding to the next prediction/observation step. Alternatively, particle filters may be viewed as a Monte Carlo integration method for the integrals involved in the state and observation equations. Isard and Blake [10] give a nice introduction to tracking using particle filters.

The SIR algorithm finds a swarm of N_p equally weighted particles which approximate the posterior $p(\boldsymbol{a}_{t-1} | X_{t-1})$ at time $t - 1$. The swarm of particles is regarded as an approximate sample from the true sample. At time t we assume that $\{\boldsymbol{a}_{t-1}^n\}_{n=1}^{N_p}$ is the swarm of N_p particles distributed as an independent sample from $p(\boldsymbol{a}_{t-1} | X_{t-1})$. Filtering proceeds as follows.

Initialisation

The filter is initialised by drawing N_p samples from the prior $p(\boldsymbol{a}_1)$. In the absence of additional information we choose $p(\boldsymbol{a}_1)$ to be Gaussian distributed with a mean found by performing static ICA on the first, say, 100 observations.

Prediction

Draw N_p samples $\{\boldsymbol{v}_t^1, \boldsymbol{v}_t^2, ... \boldsymbol{v}_t^{N_p}\}$ from the state noise density $\mathcal{N}(0, Q)$. Each particle is then propagated through the state equation 2.14 to form a new swarm of particles, $\{a^1_{t|t-1}, ..., a^{N_p}_{t|t-1}\}$, where

$$a^n_{t|t-1} = F a^n_t + v^n_t \tag{2.20}$$

If the particles $\{a^n_{t-1}\}$ are an independent sample from $p(a_{t-1} | x_{t-1})$ the particles $a^n_{t|t-1}$ are an independent sample from $p(a_t | X_{t-1})$, so the prediction stage implements equation 2.15.

The prediction stage is rapidly achieved since sampling from the Gaussian distribution is efficient.

Filtering

On the observation of a new datum x_t the prediction represented by the swarm $\{a^n_{t|t-1}\}$ can be corrected. Each particle is weighted by the likelihood of the observation x_t being generated by the mixing matrix represented by $a^n_{t|t-1}$. Thus probability masses q^n_t are assigned by

$$q^n_t = \frac{p(x_t | a^n_{t|t-1})}{\sum_{k=1}^{N_p} p(x_t | a^k_{t|t-1})} \tag{2.21}$$

This procedure can be regarded as a discrete approximation to equation 2.16, where the prior $p(a_t | X_{t-1})$ is approximated by the sample $p(a_t | X_{t-1})$.

Laplace's approximation can be used to approximate the convolution of equation 2.13 for any fixed A_t when the observational noise is small (see appendix). Otherwise the integral can be evaluated by Monte Carlo integration, however, since the integral must be evaluated for every particle, Monte Carlo integration is usually prohibitively time-consuming.

Resampling

The particles $a^n_{t|t-1}$ and weights q^n_t define a discrete distribution approximating $p(a_t | x_t)$. These are then resampled with replacement N_p times to form an approximate sample from $p(a_t | x_t)$, each particle carrying an equal weight. This new sample can now be used as the basis for the next prediction.

Although many variants and improvements to the basic SIR filter exist (see [7] for a review), we have not found the basic SIR filter to be adequate for non-stationary ICA.

2.4.1 Source Recovery

Rather than making strictly Bayesian estimates of the model parameters $\theta^m = \{r_m, w_m, \nu_m\}$, the maximum *a posteriori* (MAP) or mean estimate of A_t is used to estimate s_t, after which maximum-likelihood estimates of the parameters are found from sequences $\{s^m_\tau\}^t_{\tau=1}$. Finding maximum-likelihood

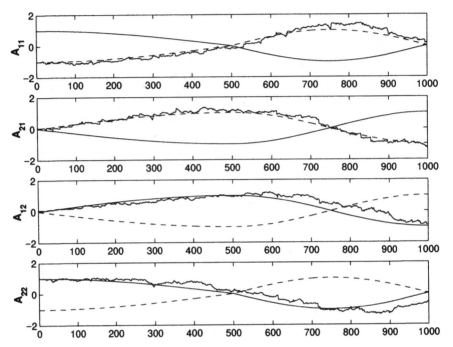

Fig. 2.2. Non-Stationary ICA tracking a mixture of Laplacian and Gaussian sources.

parameters is readily and robustly accomplished [6]. Each s_t is found by maximising $\log p(s_t | x_t, A_t)$, which is equivalent to minimising

$$(x_t - A_t^* s_t)^T R^{-1} (x_t - A_t^* s_t) + \sum_{m=1}^{M} \left| \frac{s_t^m - \nu_m}{w_m} \right|^{r_m} \qquad (2.22)$$

where A_t^* is the MAP estimate for A_t. The minimisation can be carried out with a pseudo-Newton method, for example. If the noise variance is small, $s_t \approx A_t^\dagger x_t$, where $A_t^\dagger = (A_t^T A_t)^{-1} A_t^T$ is the pseudo-inverse of A_t, and this estimate provides a good starting guess for a numerical minimisation.

2.5   Illustration of Non-Stationary ICA

Here we illustrate the method with two examples.

In the first example a Laplacian source $(p(s) \propto e^{-|s|})$ and a source with uniform density are mixed with a mixing matrix whose components vary sinusoidally with time:

$$A_t = \begin{bmatrix} \cos \omega t & \sin \omega t \\ -\sin \omega t & \cos \omega t \end{bmatrix} \qquad (2.23)$$

Note, however, that the oscillation frequency doubles during the second half of the simulation making it more difficult to track. Figure 2.2 shows the true mixing matrix and the tracking of it by non-stationary ICA. The solid line marks the mean of the particle swarm. As can be seen from the figure the particle mean tracks the mixing matrix well, although the tracking is poorer in the second half of the simulation where the mixing matrix is changing more rapidly.

Like ICA for stationary mixing, this method cannot distinguish between a column of A_t and a scaling of the column. In figure 2.2 the algorithm has "latched on" to the negative of the first column of A_t (shown dashed) which is then tracked for the rest of the simulation.

We resolve the scaling ambiguity between the variance of the sources and the scale of the columns of A_t by insisting that the variance of each source is unity; i.e.,we ignore the estimated value of w_m (equation 2.18), instead setting $w_m = 1$ for all m and allowing all the scale information to reside in the columns of A_t.

To provide an initial estimate of the mixing matrix and source parameters static ICA was run on the first 100 samples. At times $t > 100$ the generalised exponential parameters were re-estimated every 10 observations.

To illustrate the particle filter, the location of every 50th particle at every 10th timestep is shown in figure 2.3. There were 1000 particles in total. The solid line here shows the true mixing matrix.

Estimates of the tracking error are provided by the covariance of the state density, which is found from the particle swarm. In this case the true A_t lies within one standard deviation of the estimated A_t almost all the time (see figure 2.3). We remark that it appears to be more difficult to track the columns associated with light-tailed sources than heavy-tailed sources. We note, furthermore, that the Gaussian case appears to be most difficult. In figure 2.2, A_{11} and A_{21} mix the Laplacian source, and the uniform source is mixed by A_{12} and A_{22} which are tracked less well, especially during the second half of the simulation. We suspect that the difficulty in tracking columns associated with nearly Gaussian sources is due to the ambiguity between a Gaussian source and the observational noise which is assumed to be Gaussian.

It is easy to envisage situations in which the mixing matrix might briefly become singular. For example, if the microphones are positioned so that each receives the same proportions of each speaker the columns of A_t are linearly dependent and A_t is singular. In this situation A_t cannot be inverted and source estimates (equation 2.22) are very poor. To cope with this we monitor the condition number of A_t; when it is large, implying that A_t is close to singular, the source estimates are discarded for the purposes of inferring the source model parameters, $\{r_m, w_m, \mu_m\}$.

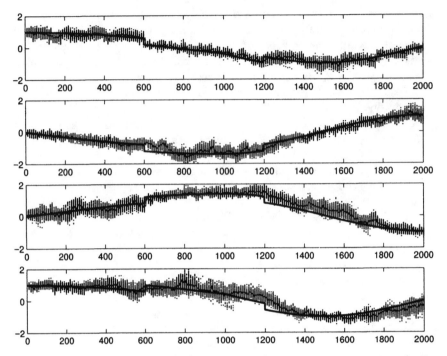

Fig. 2.3. The swarm of particles tracking the mixing matrix for a mixture of a Laplacian and a Gaussian source. Thick solid lines mark the true mixing matrix; thin solid lines mark the mean of the particle swarm. Every 50th particle is plotted every 10th timestep.

In figure 2.4 we show non-stationary ICA applied to Laplacian and uniform sources mixed with the matrices

$$A_t = \begin{bmatrix} +\cos 2\omega t & \sin \omega t \\ -\sin 2\omega t & \cos \omega t \end{bmatrix} \tag{2.24}$$

where ω is chosen so that A_{1000} is singular. Clearly the mixing matrix is tracked through the singularity, although not so closely as when A_t is well conditioned. Figure 2.5 shows the condition number of the MAP A_t. The normalising constant $Z = p(x_t | X_{t-1})$ in the correction equation 2.16 is known as the innovations probability (equation 2.17) and measures the degree to which a new datum fits the dynamic model learned by the tracker. Discrete changes of state are signalled by low innovations probability. The innovations probability is approximated by

$$p(x_t | X_{t-1}) \approx \sum_{n=1}^{N_p} p(x_t | a_{t|t-1}^n) \tag{2.25}$$

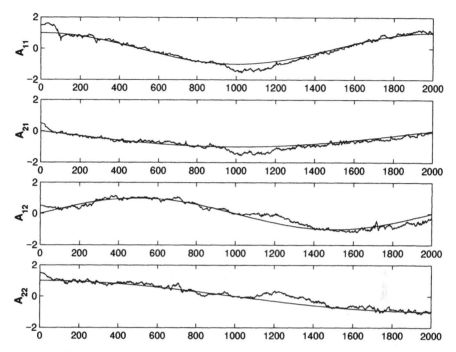

Fig. 2.4. Tracking through a singularity. The mixing matrix is singular at $t = 1000$.

when the state density is represented by particles. Figure 2.5 shows the innovations probability for the mixing shown in figure 2.4: the presence of the singularity is clearly reflected.

Note also that the simulation shown in figure 2.4 was deliberately initialised fairly close to, but not exactly at, the true A_1. The "latching on" of the tracker to the correct mixing matrix in the first 100 observations is evident in the figure.

2.6   Smoothing

The filtering methods presented estimate the mixing matrix as $p(A_t | X_t)$. They are therefore strictly causal and can be used for online tracking. If the data are analysed retrospectively future observations $(\boldsymbol{x}_\tau, \tau > t)$ may be used to refine the estimate of A_t. The Markov structure of the generative model permits the pdf $p(\boldsymbol{a}_t | X_T)$ to be found from a forward pass through the data, followed by a backward sweep in which the influence of future observations on $\boldsymbol{a}_t$ is evaluated. See, for example, [8] for a detailed exposition of forward-backward recursions. In the forward pass the joint probability

$$p(\boldsymbol{a}_t, \boldsymbol{x}_1, ... \boldsymbol{x}_t) \equiv \alpha_t = \int \alpha_{t-1} p(\boldsymbol{a}_t | \boldsymbol{a}_{t-1}) \, p(\boldsymbol{x}_t | \boldsymbol{a}_t) \, d\boldsymbol{a}_{t-1} \qquad (2.26)$$

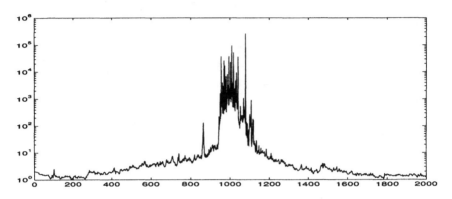

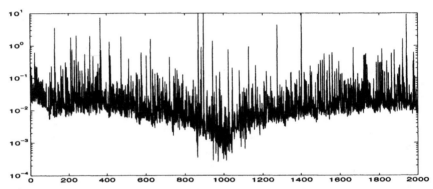

Fig. 2.5. Top: Condition number of the MAP estimate of A_t. At $t = 1000$ the true mixing matrix is singular. Matrices with condition numbers greater than 10 were not used for estimating the source parameters. Bottom: Innovations probability $p(x_t | X_{t-1})$.

is recursively evaluated. In the backward sweep the conditional probability

$$p(x_{t+1}, ..., x_T | a_t) \equiv \beta_t = \int \beta_{t+1}\, p(a_{t+1} | a_t) p(x_{t+1} | a_{t+1})\, da_{t+1} \qquad (2.27)$$

is found. Finally the two are combined to produce a smoothed non-causal estimate of the mixing matrix:

$$p(a_t | x_1, ..., x_T) \propto \alpha_t \beta_t \qquad (2.28)$$

The forward density α_t and the backward density can each be approximated by a swarm of particles. However, combining them to produce the smoothed estimate (equation 2.28) necessitates storing the entire history of the particles during the forward sweep [14]. The storage problems inherent in this method can be somewhat alleviated by ignoring the influence of observations outside some window around the current observation. An alternative,

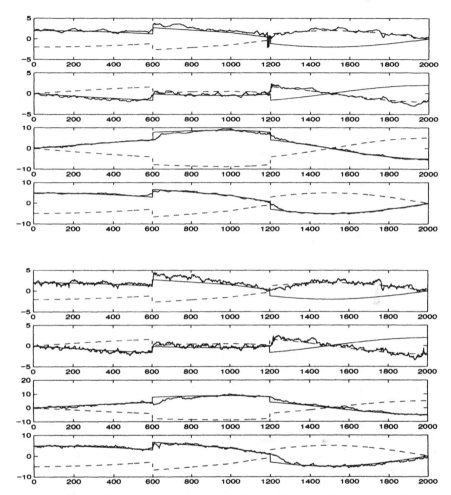

Fig. 2.6. Top: Retrospective tracking with forward-backward recursions. Bottom: Online filtering of the same data. Dashed lines show the negative of the mixing matrix elements.

which we have adopted is to calculate the mean and the covariance of the particle swarm approximating α_t at each t on the forward sweep. These are then combined with a Gaussian approximation to β_t from the backward sweep. The Gaussian approximation makes the calculation of equation 2.28 particularly simple and only two quantities, the mean and the covariance of α_t, need be stored. Figure 2.6 illustrates tracking by both smoothing and causal filtering. As before the elements of the mixing matrix vary sinusoidally with time except for discontinuous jumps at $t = 600$ and 1200. Both the filtering and forward-backward recursions track the mixing matrix; however the smoothed

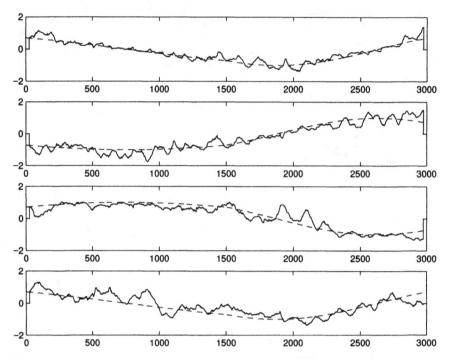

Fig. 2.7. Tracking temporally correlated sources. Elements of the mixing matrix during tracking of temporally correlated sources. Dashed lines show the true mixing matrix elements.

estimate is less noisy and more accurate, particularly at the discontinuities. Note also that, following the discontinuity at $t = 1200$, the negative of the first columm of A_t is tracked.

2.7   Temporal Correlations

The graphical model in figure 2.1 assumes that successive samples from each source are independent, so that the sources are stochastic. When temporal correlations in the sources are present the model must be modified to include the conditional dependence of s_t^m on s_{t-1}^m. In this case the hidden state is now comprised of a_t and the states of the sources s_t, and predictions and corrections for the full state should be made. Since the sources are independent, predictions for each source and a_t may be made independently and the system is a factorial hidden Markov model [8].

A number of source predictors have been implemented, including the Kalman filter, AR models and Gaussian mixture models. However, we find that in all these cases the combined tracker is unstable. The instability arises because the change in observation from x_{t+1} to x_t cannot be unambiguously

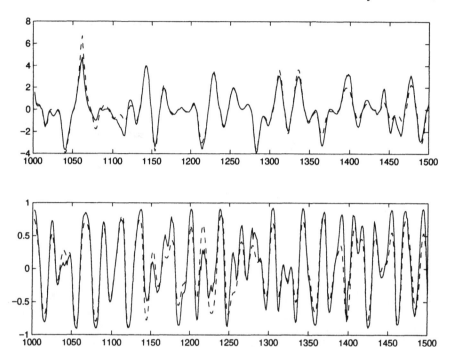

Fig. 2.8. Recovered sources and the true sources (dashed) for times 1000-1500 for the tracking shown in figure 2.7

assigned to either a change in the mixing matrix or a change in the sources. Small errors in the prediction of the sources induce errors in the mixing matrix estimates, which in turn lead to errors in subsequent source predictions; these errors are then incorporated into the predictive model for the sources and further (worse) errors in the prediction are made. This problem is not present in the stochastic case because the source model is much more tightly constrained.

Under the assumption that the sources evolve on a rapid timescale compared with the mixing matrix, the effect of temporal correlations in the sources may be removed by averaging over a sliding window. That is, the likelihood $p(\boldsymbol{x}_t \,|\, A_t)$ used in the correction step (equation 2.16) is replaced by

$$\left\{ \prod_{\tau=-L}^{L} p(\boldsymbol{x}_{t+\tau} \,|\, A_{t+\tau}) \right\}^{\frac{1}{2L+1}} \tag{2.29}$$

The length of the window $2L + 1$ is chosen to be of a typical timescale of the sources. Tracking using the averaged likelihood is computationally expensive because at each t the $p(\boldsymbol{x}_{t+\tau} \,|\, A_{t+\tau})$ must be evaluated for each τ in the sliding window. An alternative method of destroying the source temporal correlations is to replace the likelihood $p(x_t | A_t)$ with $p(x_{t+\tau} | A_{t+\tau})$ with τ chosen at random from within the sliding window $(-L \leq \tau \leq L)$. This is no more expensive than using $p(x_t | A_t)$ and effectively destroys the source correlations.

Figures 2.7 and 2.8 illustrate the tracking of a mixing matrix with temporally correlated sources. The window length was $L = 50$. Tracking is not as accurate as in the stochastic case, however the mixing matrix is followed and the sources are recovered well.

If more precise knowledge about the sources is available it might be incorporated into the source evolution equations in the form of a Bayesian prior. With sufficient information to constrain the sources this might permit the tracking and separation of temporally correlated sources without recourse to temporal averaging.

2.8   Conclusion

We have presented a method for independent component analysis when the mixing proportions are non-stationary. The Sampling Importance Resampling particle filter permits easy tracking of the state density even though the density is non-Gaussian and the observation equation is nonlinear. The method is strictly causal and can be used for online tracking (or "filtering"). If data are analysed retrospectively forward-backward recursions may be used for smoothing rather than filtering. The mixing of temporally correlated sources may be tracked by averaging or sampling from within a sliding window.

In common with most tracking methods, the state noise covariance Q and the observational noise covariance R are parameters which must be set. Although we have not addressed the issue here, it is straightforward, though laborious, to obtain maximum-likelihood estimates for them using the EM method [8]. It would also be possible to estimate the state mixing matrix F in the same manner.

Although we have modelled the source densities here with generalised exponentials, which permits the separation of a wide range of sources, it is possible to generalise or restrict the source model. More complicated (possibly multi-modal) densities may be represented by a mixture of Gaussians. On the other hand, if all the sources are restricted to be Gaussian the method becomes a tracking factor analyser. In the zero noise limit the method performs non-stationary principal components analysis.

2.8.1   Acknowledgement

We gratefully acknowledge funding from British Aerospace plc.

2.9 Appendix: Laplace's Approximation for the Likelihood

Here we use Laplace's approximation to evaluate the convolution integral equation 2.13 which gives the likelihood of an observation x given the mixing matrix A:

$$p(x|A) = \int \mathcal{G}(x - As, R) \prod_{m=1}^{M} p_m(s^m) \, ds \tag{2.30}$$

The source densities are taken to be generalised exponentials (equation 2.18).

Laplace's approximation [18] for the N-dimensional integral

$$I = \int f(y) \exp\{-h(y)\} \, dy \tag{2.31}$$

where $h(y)$ is sharply peaked, is given by

$$I \approx \int f(\hat{y}) \exp\{-h(\hat{y}) - (y - \hat{y})^T H(y - \hat{y})/2\} \, dy \tag{2.32}$$

Here $H = \frac{\partial^2 h}{\partial y_i y_j}$ is the Hessian and $\hat{y}$ is the y that minimises h. Thus

$$I \approx f(\hat{y}) \exp -h(\hat{y}) \sqrt{\det 2\pi H^{-1}} \tag{2.33}$$

$$= \frac{(2\pi)^{\frac{N}{2}}}{\sqrt{\det H}} f(\hat{y}) \exp -h(\hat{y}) \tag{2.34}$$

If the observational noise is small compared to the amplitude of the sources we regard the Gaussian term as $\exp\{-h\}$ and the generalised exponential and normalising factors as f. Then $h(s) = (As - x)^T R^{-1}(Ax - x)/2$ is minimised at $\hat{s} = (A^T C^{-1} A)^{-1} A^T R^{-1} x$ and the Hessian is $H = A^T R^{-1} A$. Consequently the likelihood is approximated as

$$p(x|A) \approx \rho \exp\{-(A\hat{s} - x)^T R^{-1}(A\hat{s} - x)/2\} \exp - \sum \left| \frac{\hat{s}_i - \nu_i}{w_i} \right|^{R_i} \tag{2.35}$$

where the normalising factor ρ is

$$\rho = \frac{Z}{\sqrt{\det(A^T R^{-1} A) \det(R)}} \tag{2.36}$$

Comparison with Monte-Carlo integration shows that this approximation is good over a wide range of noise covariances. Evaluation of the likelihood using equation 2.35 is much more efficient than Monte-Carlo integration or other numerical quadratures.

References

1. S. Amari, A. Cichocki, and H. Yang. A new learning algorithm for blind signal separation. In D. Touretzky, M. Mozer, and M. Hasselmo, editors, *Advances in Neural Information Processing Systems*, **8**, 757–763, MIT Press, Cambridge MA, 1996.

2. H. Attias. Independent factor analysis. *Neural Computation*, **11**(5):803–852, 1999.

3. A. J. Bell and T. J. Sejnowski. An information-maximization approach to blind separation and blind deconvolution. *Neural Computation*, **7**(6):1129–1159, 1995.

4. J-F. Cardoso. Infomax and maximum likelihood for blind separation. *IEEE Sig. Proc. Letters*, **4**(4):112–114, 1997.

5. J-F. Cardoso. On the stability of source separation algorithms. In T. Constantinides, S.-Y. Kung, M. Niranjan, and E. Wilson, editors, *Neural Networks for Signal Processing VIII*, 13–22. IEEE Signal Processing Society, IEEE, 1998.

6. R. M. Everson and S. J. Roberts. Independent Component Analysis: A flexible non-linearity and decorrelating manifold approach. *Neural Computation*, **11**(8), 1999. Available from http://www.dcs.ex.ac.uk/academics/reverson.

7. P. Fearnhead. *Sequential Monte Carlo methods in filter theory*. PhD thesis, Merton College, University of Oxford, 1999.

8. Z. Ghahramani. Learning dynamic Bayesian networks. In C. L. Giles and M. Gori, editors, *Adaptive Processing of Temporal Information*, Lecture Notes in Artificial Intelligence. Springer-Verlag, 1999.

9. N. Gordon, D. Salmond, and A. F. M. Smith. Novel approach to nonlinear/non-Gaussian Bayesian state estimation. *IEE Proceedings-F*, **140**:107–113, 1993.

10. M. Isard and A. Blake. Contour tracking by stochastic density propagation of conditional density. In *Proc. European Conf. Computer Vision*, 343–356. Cambridge, UK, 1996.

11. A. H. Jazwinski. *Stochastic Processes and Filtering Theory*. Academic Press, 1973.

12. I. T. Joliffe. *Principal component analysis*. Springer-Verlag, 1986.

13. R. Kalman and R. Bucy. New results in linear filtering and prediction theory. *Journal of Basic Engineering, Trans. ASME Series D*, **83**(95 - 108), 1961.

14. G. Kitigawa. Monte Carlo filter and smoother for non-Gaussian nonlinear state space models. *Journal of Computational and Graphical Statistics*, 5:1–25, 1996.

15. T-W. Lee, M. Girolami, A. J. Bell, and T. J. Sejnowski. A unifying information-theoretic framework for independent component analysis. *International Journal on Mathematical and Computer Modeling*, 1998. (In press). Available from http://www.cnl.salk.edu/~tewon/Public/mcm.ps.gz.

16. T-W. Lee, M. Girolami, and T. J. Sejnowski. Independent component analysis using an extended infomax algorithm for mixed sub-Gaussian and super-Gaussian sources. *Neural Computation*, **11**:417–441, 1999.

17. D. J. C. MacKay. Maximum likelihood and covariant algorithms for independent component analysis. Technical report, University of Cambridge, December 1996. Available from http://wol.ra.phy.cam.ac.uk/mackay/.

18. J. J. K. O' Ruanaidth and W. J. Fitzgerald. *Numerical Bayesian Methods Applied to Signal Processing*. Springer, 1996.

19. A. Papoulis. *Probability, Random Variables and Stochastic Processes.* McGraw-Hill, 1991.

20. B. Pearlmutter and L. Parra. A context-sensitive generalization of ICA. In *International Conference on Neural Information Processing*, 1996.

Part II

The Validity of the Independence Assumption

3 The Independence Assumption: Analyzing the Independence of the Components by Topography

Aapo Hyvärinen, Patrik O. Hoyer and Mika Inki

Summary. In ordinary independent component analysis, the components are assumed to be completely independent, and they do not necessarily have any meaningful order relationships. In practice, however, the estimated "independent" components are often not at all independent. We propose that this residual dependence structure could be used to define a topographic order for the components. In particular, a distance between two components could be defined using their higher-order correlations, and this distance could be used to create a topographic representation. Thus we obtain a linear decomposition into approximately independent components, where the dependence of two components is approximated by the proximity of the components in the topographic representation.

3.1 Introduction

Independent component analysis (ICA) [19] is a statistical model where the observed data is expressed as a linear transformation of latent variables that are non-Gaussian and mutually independent. The classic version of the model can be expressed as

$$\mathbf{x} = \mathbf{As} \tag{3.1}$$

where $\mathbf{x} = (x_1, x_2, ..., x_n)^T$ is the vector of observed random variables, $\mathbf{s} = (s_1, s_2, ..., s_n)^T$ is the vector of the independent latent variables (the "independent components"), and $\mathbf{A}$ is an unknown constant matrix, called the mixing matrix. The problem is then to estimate both the mixing matrix $\mathbf{A}$ and the realizations of the latent variables s_i, using observations of $\mathbf{x}$ alone. Exact conditions for the identifiability of the model were given in [9]; the most fundamental is that the independent components s_i must be nongaussian [9]. A considerable amount of research has been recently conducted on the estimation of this model, see e.g. [1,3,7,6,8,10,4,12,20,26,29].

In classic ICA, the independent components s_i have no particular order, or other relationships. It is possible, though, to define an order relation between the independent components by such criteria as non-Gaussianity or contribution to the observed variance [14]; the latter is given by the norms of the corresponding columns of the mixing matrix as the independent components are defined to have unit variance. Such trivial order relations may be useful for some purposes, but they are not very informative in general.

The lack of an inherent order of independent components is related to the assumption of complete statistical independence. In practical applications of ICA, however, one can very often observe clear violations of the independence assumption. It is possible to find, for example, pairs of estimated independent components that are clearly dependent on each other. This dependence structure is often very informative, and it would be useful to estimate it somehow.

Estimation of the "residual" dependency structure of estimates of independent components could be based, for example, on computing the cross-cumulants. Typically these would be higher-order cumulants since second-order cross-cumulants, i.e. the covariances, are typically very small, and are in fact forced to be zero in many ICA estimation methods, e.g. [9,4,12]. A more information-theoretic measure for dependence would be given by mutual information. Whatever measure is used, however, the problem remains as to how such numerical estimates of the dependence structure should be visualized or otherwise utilized. Moreover, there is another serious problem associated with simple estimation of some dependency measures from the estimates of the independent components. This is due to the fact that often the independent components do not form a well-defined set. Especially in image decomposition [3,27,28,13], the set of potential independent components seems to be larger than what can be estimated at one time, in fact the set might be infinite. A classic ICA method gives an arbitrarily chosen subset of such independent components. Thus, it is important in many applications that the dependency information is utilized during the estimation of the independent components, so that the estimated set of independent components is one that can be ordered in a meaningful way.

We propose here that the residual dependency structure of the "independent" components, i.e. dependencies that cannot be cancelled by ICA, could be used to define a *topographic order* between the components. The topographic order is easy to represent by visualization, and has the usual computational advantages associated with topographic maps [11,23,33]. We propose a modification of the ICA model that explicitly formalizes a topographic order between the independent components. This gives a topographic map where the distance of the components in the topographic representation is a function of the dependencies of the components. Components that are near to each other in the topographic representation are strongly dependent in the sense of higher-order correlations, or mutual information. This gives a new principle for topographic organization. Furthermore, we derive a learning rule for the estimation of the model. Finally, we show experiments to validate our approach, including feature extraction from image and audio data, and blind separation of magnetoencephalographic components.

3.2   Background: Independent Subspace Analysis

In classic ICA, all the components s_i are assumed independent. For future reference, let us consider the likelihood of the model. Denoting by $\mathbf{W} = (\mathbf{w}_1, ..., \mathbf{w}_n)^T$ the matrix $\mathbf{A}^{-1}$, we can write the log - likelihood of the model, given T observations $\mathbf{x}(t), t = 1, ..., T$, as follows [30]:

$$\log L(\mathbf{w}_i, i = 1, ..., n) = \sum_{t=1}^{T} \sum_{i=1}^{n} \log p_i(\mathbf{w}_i^T \mathbf{x}(t)) + T \log |\det \mathbf{W}| \qquad (3.2)$$

where the p_i are the density functions of the s_i (here assumed to be known).

In [16], a modification of the ICA model was introduced, in which the components s_i were not assumed to be mutually independent. This model, called independent subspace analysis, is the starting point of topographic ICA.

The inspiration for independent subspace analysis comes from coding invariant features. The classical approach of feature extraction is to use linear transformations, or filters. The presence of a given feature is detected by computing the dot-product of input data with a given feature vector. For example, wavelet, Gabor, and Fourier transforms, as well as most models of V1 simple cells, use such linear features. The problem with linear features is, however, that they necessarily lack any invariance with respect to such transformations as spatial shift or change in (local) Fourier phase [31,24]. Kohonen [24] developed the principle of invariant-feature subspaces as an abstract approach to representing features with some invariances. The principle of invariant-feature subspaces states that one may consider an invariant feature as a linear subspace in a feature space. The value of the invariant, higher-order feature is given by (the square of) *the norm of the projection* of the given data point on that subspace, which is typically spanned by lower-order features.

To link invariant-feature subspaces with ICA, we have to relax the assumption of the independence of all the components s_i. Instead, in [16] it was assumed that the s_i can be divided into couples, triplets or in general m-tuples, such that the s_i inside a given m-tuple could be dependent on each other, but dependencies between different m-tuples were not allowed. Related relaxations of the independence assumption were proposed in [5,25].

Inspired by the principle of feature subspaces [24], the probability densities for the m-tuples of s_i were assumed in [16] to be *spherically symmetric*, i.e. to depend only on the norm. In other words, the probability density $p_q(.)$ of the m-tuple with index $q \in \{1, ..., Q\}$, could be expressed as a function of the sum of the squares of the $s_i, i \in S_q$ only, where we denote by S_q the set of indices of the components s_i that belong to the q-th m-tuple. (For simplicity, it was assumed further that the $p_q(.)$ were equal for all q, i.e. for all subspaces.) In this model of independent subspace analysis, the logarithm of the likelihood can be expressed as

$$\log L(\mathbf{w}_i, i = 1, ..., n) = \sum_{t=1}^{T} \sum_{q=1}^{Q} G(\sum_{i \in S_q} (\mathbf{w}_i^T \mathbf{x}(t))^2) + T \log |\det \mathbf{W}| \quad (3.3)$$

where $G(\sum_{i \in S_q} s_i^2) = \log p_q(s_i, i \in S_q)$ gives the logarithm of the probability density inside the q-th m-tuple of s_i.

The independent subspace model introduces a certain dependence structure for the independent components. Let us assume that the distribution in the subspace is sparse, which means that the norm of the projection is most of the time very near to zero; in other words, the norm has a distribution similar to the distribution of the absolute value of a super-Gaussian variable [4]. For example, we could use the following function [16]

$$G(\sum_{i \in S_j} s_i^2) = -\alpha[\sum_{i \in S_j} s_i^2]^{1/2} + \beta, \quad (3.4)$$

which could be considered a multi-dimensional version of the exponential distribution.

Then the model implies that two components s_i and s_j that belong to the same subspace tend to be non-zero simultaneously. This seems to be a preponderant structure of dependency in most natural data. For image data, this has also been noted by Simoncelli [32]. Such a dependency is given by a certain kind of higher-order correlation, namely correlation of energies. This means that

$$\text{cov } (s_i^2, s_j^2) = E\{s_i^2 s_j^2\} - E\{s_i^2\}E\{s_j^2\} > 0 \quad (3.5)$$

if s_i and s_j are close in the topography. Note that although the components tend to be active, i.e. non-zero, at the same time, the actual values of s_i and s_j are not easily predictable from each other.

For example, if the variables are defined as products of two zero-mean independent components z_i, z_j and a common "variance" variable σ:

$$s_i = z_i \sigma$$
$$s_j = z_j \sigma \quad (3.6)$$

then s_i and s_j are uncorrelated, but their energies are not. In fact the covariance of their energies equals $E\{z_i^2 \sigma^2 z_j^2 \sigma^2\} - E\{z_i^2 \sigma^2\}E\{z_j^2 \sigma^2\} = E\{\sigma^4\} - E\{\sigma^2\}^2$, which is positive because it equals the variance of σ^2 (we assumed here for simplicity that z_i and z_j are of unit variance). This kind of dependence is illustrated in figure 3.1.

Moreover, we see from equation 3.3 that choosing a sparse log-density G implies that the estimation of the model consists of finding subspaces such that the norms of the projections of the data on those subspaces have maximally sparse distributions. According to the principle of invariant feature subspaces, it is to be expected that the norms of the projections on the subspaces represent some higher-order, invariant features. The exact nature of the invariances has not been specified in the model but will emerge from the input data, using only the prior information on their independence.

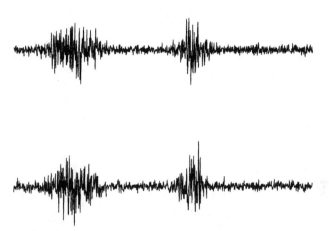

Fig. 3.1. Illustration of higher-order dependencies. The two signals in the figure are uncorrelated but they are not independent. In particular, their energies are correlated. The signals were generated as in equation 3.6, but for purposes of illustration, the variable σ was replaced by a time-correlated signal.

3.3   Topographic ICA Model

3.3.1   Dependence and Topography

In this section, we define topographic ICA by modifying slightly the model of independent subspace analysis as described in the preceding section. An alternative formulation using an explicit generative model that is a hierarchical version of the ordinary ICA model is given in section 3.3.3 below and in [15].

The idea is to relax the assumption of the independence of the components s_i in equation 3.1 so that components that are close to each other in the topography are *not* assumed to be independent in the model. In contrast, components that are not close to each other in the topography *are* independent, at least approximately; thus most pairs of components are independent. (Of course, if independence does not hold for most component pairs, any connection to ICA is lost, and the model would not be very useful in those applications where ICA has proved useful.) Thus, if the topography is defined by a lattice or grid, the dependency of the components is a function of the distance of the components on that grid.

The basic problem is then to *choose what kind of dependencies are allowed* between nearby components. The most basic dependence relation is linear correlation.[1] However, allowing linear correlation between the components does not seem very useful. In fact, in many ICA estimation methods, the components are constrained to be uncorrelated [7,9,4,12], so the requirement not to be correlated seems natural in any extension of ICA as well.

In fact, according to the arguments of the preceding section, a very important kind of dependency is given by correlation of energies. This is modelled by independent subspace analysis, although only inside each subspace. So, generalizing that model, we could define a model that models topographic correlations of energies. This is the approach we choose here.

3.3.2 Defining Topographic ICA

Now we generalize the model defined by equation 3.3 so that it models a dependence not only inside the m-tuples, but among all neighbouring components. A neighbourhood relation defines a topographic order [23]. We define the model by the following likelihood:

$$\log L(\mathbf{w}_i, i = 1, ..., n) = \sum_{t=1}^{T}\sum_{j=1}^{n} G(\sum_{i=1}^{n} h(i,j)(\mathbf{w}_i^T \mathbf{x}(t))^2) + T \log |\det \mathbf{W}|$$

(3.7)

Here, the $h(i,j)$ is a neighbourhood function, which expresses the strength of the connection between the i-th and j-th units. The neighbourhood function can be defined in the same way as with the self-organizing map [23]. Neighbourhoods can thus be defined as one-dimensional or two-dimensional; 2-D neighbourhoods can be square or hexagonal. Usually, the neighbourhood function is symmetric: $h(i,j) = h(j,i)$. A simple example is to define a 1-D neighbourhood relation by

$$h(i,j) = \begin{cases} 1, & \text{if } |i - j| \leq m \\ 0, & \text{otherwise.} \end{cases}$$

(3.8)

The constant m defines here the width of the neighbourhood: The neighbourhood of the component with index i consists of those components whose indices are in the range $i - m, ..., i + m$. For the Kronecker delta function, $h(i,j) = \delta_{ij}$, which corresponds to $m = 0$, the model reduces to classic ICA.

The function G has a similar role to that of the log-density of the independent components in classic ICA. For image data, or other data with a

---

[1] In this paper, we mean by correlation a normalized form of covariance: $\text{corr}(s_1, s_2) = [E\{s_1 s_2\} - E\{s_1\}E\{s_2\}][\text{var}(s_1)\text{var}(s_2)]^{-1/2}$.

sparse structure, G should be chosen so that it corresponds to the log-density of the square of a sparse variable. For example, one could take as in [16]:

$$G_1(u) = -\alpha_1 \sqrt{u} + \beta_1 \tag{3.9}$$

which could be considered a multi-dimensional version of the exponential distribution. The scaling constant α_1 and the normalization constant β_1 are determined so as to give a probability density that is compatible with the constraint of unit variance of the s_i, but they are irrelevant in the following. An alternative would be to use polynomial approximation:

$$G_2(u) = \alpha_2 u^2 + \beta_2 \tag{3.10}$$

although this does not give a properly defined density.

Our model can thus be considered a generalization of the model of independent subspace analysis [16]. In independent subspace analysis, the components s_i are divided into m-tuples or subspaces, so that the component s_i belongs to the neighbourhood of the component s_j if and only if they belong to the same m-tuple. This could be expressed as a very special case of topographic ICA with a neighbourhood function of the form

$$h(i,j) = \begin{cases} 1, & \text{if } \exists q : i,j \in S_q \\ 0, & \text{otherwise} \end{cases} \tag{3.11}$$

In topographic ICA, such subspaces are completely overlapping: every component has its "own" neighbourhood, and could be considered to define a subspace. Thus each of the terms $G(\sum_{i=1}^{n} h(i,j)(\mathbf{w}_i^T \mathbf{x})^2)$ could be considered as a (weighted) projection on a feature subspace, i.e. as the value of an invariant feature.

3.3.3 The Generative Model

An alternative approach is to define the distribution of the components s_i using an explicit generative model.

In this approach, we generate the s as follows. The variances σ_i^2 of the s_i are not constant, instead they are assumed to be random variables, generated according to a model to be specified. After generating the variances, the variables s_i are generated independently from each other, using some conditional distributions to be specified. In other words, the s_i are *independent given their variances*. Dependence among the s_i is implied by the dependence of their variances. According to the principle of topography, the variances corresponding to nearby components should be (positively) correlated, and the variances of components that are not close should be independent, at least approximatively.

Using the topographic relation $h(i, j)$, many different models for the variances σ_i^2 could be used. We prefer here to define them by an ICA model followed by a nonlinearity:

$$\sigma_i = \phi(\sum_{k=1}^{n} h(i, k)u_k) \tag{3.12}$$

where u_i are the "higher-order" independent components used to generate the variances and ϕ is some scalar nonlinearity. The distributions of the u_i and the actual form of ϕ are additional hyperparameters of the model.

The resulting generative model is summarized in figure 3.2. Note that the two stages of the generative model can be expressed as a single equation as follows:

$$s_i = \phi(\sum_{k} h(i, k)u_k)z_i \tag{3.13}$$

where z_i is a random variable that has the same distribution as s_i given that σ_i^2 is fixed to unity. The u_i and the z_i are all mutually independent.

The definition given in section 3.3.2 can be considered as an approximation of the explicit generative model given here. This is useful because the likelihood of the former definition is given in closed form, whereas the likelihood of the explicit generative model cannot be expressed analytically. For details, see [15].

3.3.4   Basic Properties of the Topographic ICA Model

Here we discuss some basic properties of the generative model defined above. Proofs of these properties can be found in [15].

1. All the components s_i are uncorrelated. To simplify things, one can define that the marginal variances (i.e. integrated over the distibution of σ_i) of the s_i are equal to unity, as in ordinary ICA. Thus the vector s can be considered to be sphered, i.e. white.
2. Components that are far from each other are more or less independent. More precisely, assume that s_i and s_j are such that their neighbourhoods have no overlap, i.e. there is no index k such that both $h(i, k)$ and $h(j, k)$ are non-zero. Then the components s_i and s_j are independent.
3. Components s_i and s_j that are near to each other, i.e. such that $h(i, j)$ is significantly non-zero, tend to be active (non-zero) at the same time. In other words, their energies s_i^2 and s_j^2 are usually positively correlated.
4. An interesting special case of topographic ICA is obtained when every component s_i is assumed to have a Gaussian distribution when the variance is given. This means that the marginal, unconditional distributions of the components s_i are continuous mixtures of gaussians. In fact, these distributions are always super-Gaussian, i.e. have positive kurtosis.

Since most independent components encountered in real data are super-Gaussian, it seems realistic to use a Gaussian conditional distribution for the s_i.

5. Classic ICA is obtained as a special case of the topographic model, by taking a neighbourhood function $h(i, j)$ that is equal to the Kronecker delta function, $h(i, j) = \delta_{ij}$.

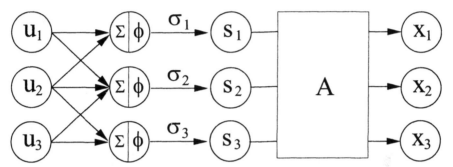

Fig. 3.2. The explicit generative model for topographic ICA. First, the "variance-generating" variables u_i are generated randomly. They are then mixed linearly inside their topographic neighbourhoods. The mixtures are then transformed using a nonlinearity ϕ, thus giving the local variances σ_i. Components s_i are then generated with variances σ_i. Finally, the components s_i are mixed linearly to give the observed variables x_i.

3.4 Learning Rule

In this section, we derive a learning rule for performing the maximization of the likelihood defined in section 3.3.2. This is possible by simple gradient learning.

First, we assume here that the data is preprocessed by whitening and that the $\mathbf{w}_i^T$, i.e. the estimates of the rows of the inverse of $\mathbf{A}$, are constrained to form an orthonormal system [9,4,12,6]. This implies that the estimates of the components are uncorrelated. Such a simplification is widely used in ICA, and it is especially useful here since it allows us to concentrate on higher-order correlations.

Thus we can simply derive a gradient algorithm in which the i-th (weight) vector $\mathbf{w}_i$ is updated as

$$\Delta \mathbf{w}_i \propto E\{\mathbf{x}(\mathbf{w}_i^T \mathbf{x}) r_i)\} \tag{3.14}$$

where

$$r_i = \sum_{k=1}^{n} h(i, k) g\left(\sum_{j=1}^{n} h(k, j)(\mathbf{w}_j^T \mathbf{x})^2\right). \tag{3.15}$$

The function g is the derivative of G, defined as in equation 3.9 or equation 3.10. After every step of equation 3.14, the vectors $\mathbf{w}_i$ are normalized to unit variance and orthogonalized, as in [4,12], for example. Note that, rigorously speaking, the expectation in (3.14) should of course be the sample average, but for simplicity, we use this notation. Of course, a stochastic gradient method could be used as well, which means omitting the averaging and taking only one sample point at a time.

In a neural interpretation, the learning rule in equation 3.14 can be considered as "modulated" Hebbian learning, since the learning term is modulated by the term r_i. This term could be considered as top-down feedback as in [16], since it is a function of the local energies which could be the outputs of higher-order neurons (complex cells).

3.5   Comparison with Other Topographic Mappings

Our method is different from ordinary topographic mappings in several ways.

The first minor difference is that whereas in most topographic mappings a single weight vector represents a single point in the data space, every vector in topographic ICA represents a direction, i.e. a one-dimensional subspace. This difference is not of much consequence, however. For example, there are versions of the Self-Organizing Map (SOM), see [22,23], that use a single weight vector in much the same way as topographic ICA.

Second, since topographic ICA is a modification of ICA, it still attempts to find a decomposition into components that are independent. This is because most components are independent in the model, at least approximately. In contrast, most topographic mappings choose the representation vectors by principles similar to vector quantization and clustering [22,23,21,4,11,33].

Most interestingly, the very principle defining topography is different in topographic ICA from that in most topographic maps. Usually, the similarity of vectors in the data space is defined by Euclidean geometry: either the Euclidean distance, as in the SOM and the GTM, or the dot-product, as in the "dot-product SOM" [23]. In topographic ICA, the similarity of two vectors in the data space is defined by their higher-order correlations, which cannot be expressed as Euclidean relations. It can be expressed using the general framework developed in [11].

In fact, the topographic similarity defined in topographic ICA could be seen as a higher-order version of the dot-product measure. If the data is prewhitened, the dot-product in the data space is equivalent to correlation in the original space. Thus, topography based on dot-products could be used to define a "second-order" topography, where components near to each other in the topography have larger linear correlations. As explained above, one can constrain the components to be uncorrelated in ICA and thus also in topographic ICA. Then any statistical dependency that could be used to create the topographic organization must be obtained from higher-order correlations and this is exactly what happens in topographic ICA.

3.6 Experiments

3.6.1 Experiments in Feature Extraction of Image Data

A very interesting application of topographic ICA can be found with image data. The data was obtained by taking 16×16 pixel image patches at random locations from monochrome photographs depicting wildlife scenes (animals, meadows, forests, etc.). The images were taken directly from PhotoCDs, and are available on the World Wide Web.[2] The mean gray-scale value of each image patch (i.e. the DC component) was subtracted. The data was then low-pass filtered by reducing the dimension of the data vector by principal component analysis, retaining the 160 principal components with the largest variances, after which the data was whitened by normalizing the variances of the principal components. These preprocessing steps are essentially similar to those used in [16,27,34]. In the results shown above, the inverse of these preprocessing steps was performed. The fact that the data was contained in a 160 dimensional subspace meant that the 160 basis vectors now formed an orthonormal system for that subspace and not for the original space, but this did not necessitate any changes in the learning rule.

The neighbourhood function was defined so that every neighbourhood consisted of a 3×3 square of 9 units on a 2-D torus lattice [23]. In other words, we defined the neighbourhood function $h(i, j)$ so that it equals one if the components i and j are adjacent in the 2-D lattice, even obliquely; otherwise, it equals zero. Experimenting with different 2-D neighbourhoods, we saw that the model is fairly robust with respect to the specification of the neighbourhood, as has also been noted in connection to Kohonen's self-organizing map [23]. The choice of a two-dimensional grid was here motivated by convenience of visualization only; further research is needed to see what the "intrinsic dimensionality" of natural image data could be. Moreover, the function G was chosen as in (3.9).

The approximation of likelihood in equation 3.7 for 50,000 observations was maximised under the constraint of orthonormality of the filters in the whitened space, using the gradient method equation 3.14.

The obtained basis vectors, i.e. columns of the mixing matrix, are shown in figure 3.3. The basis vectors are similar to those obtained by ordinary ICA of image data [27,3]. In addition, they have a clear topographic organisation.

The topographic organization was investigated in more detail by computing the correlations of the energies of the components. In figure 3.4, the correlations of the energies are plotted as a function of the distance on the

---

[2] http://www.cis.hut.fi/projects/ica/data/images/

topographic grid. One can see that the correlations are decreasing as the distance increases. This was predicted by the model. After a certain distance, however, the correlations no longer decrease, reaching a constant value. According to the model, the correlations should continue decreasing and reach zero, but this does not happen exactly because image data does not exactly follow the model. It is probable, however, that for a much larger window size, the correlations would go to zero.

The connection to independent subspace analysis [16], which is basically a complex cell model, can also be found in these results. Two neighbouring basis vectors in figure 3.3 tend to be of the same orientation and frequency. Their locations are also near to each other. In contrast, their phases are very different. This means that a neighbourhood of such basis vectors, i.e. simple cells, functions as a complex cell: The local energies that are summed in the approximation of the likelihood in equation 3.7 can be considered as the outputs of a complex cell. Likewise, the feedback r_i in the learning rule could be considered as coming from complex cells. In [15], the complex cell interpretation was investigated in more detail using the same methods as in [16].

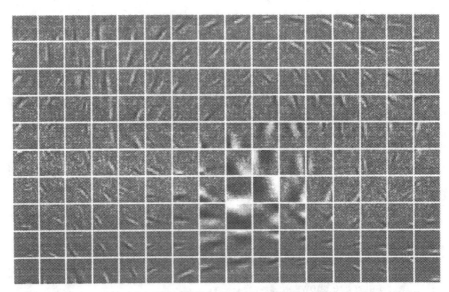

Fig. 3.3. Topographic ICA of natural image data. The model gives Gabor-like basis vectors for image windows. Basis vectors that are similar in location, orientation or frequency are close to each other. The phases of nearby basis vectors are very different, giving each neighbourhood properties similar to those of complex cells.

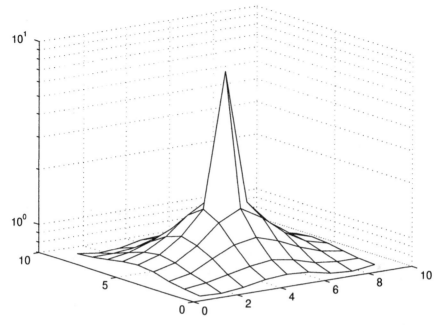

Fig. 3.4. Analysis of the higher-order correlations of the components estimated from image data. The plot shows the covariances of energies (in log-scale) of the components as a function of the relative position on the topographic grid. The covariances were averaged over all components. The plot shows that the covariances are a decreasing function of distance.

3.6.2   Experiments in Feature Extraction of Audio Data

In the second set of experiments, we used speech data for feature extraction. The data set consisted of excerpts of speech collected from various speakers. The excerpts (sampled at 16 kHz and 16 bits) were concatenated and fragments of length 256 were randomly sampled from the resulting audio file. These formed the 256-dimensional input data vectors, which were first reduced in dimension to 100 with PCA, in order to remove noise and uninteresting signals. We then estimated the topographic ICA model from the data using again a two-dimensional torus-shaped map (10 × 10 units).

The results are shown in figure 3.5. These features seem to be organized according to frequency and phase, although again no such information was explicitly used in the algorithm.

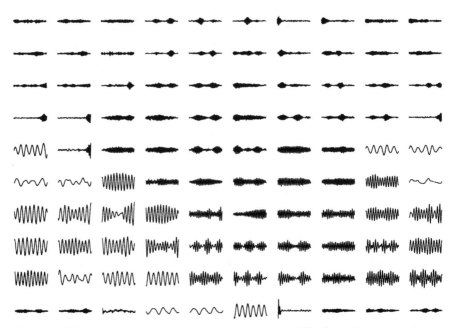

Fig. 3.5. The range of features found by topographic ICA from the speech data.

3.6.3 Experiments with Magnetoencephalographic Recordings

Next we did experiments on blind source separation. Two minutes of magnetoencephalographic (MEG) data was collected using a 122-channel whole-scalp neuromagnetometer device. The sensors measured the gradient of the magnetic field in two orthogonal directions at 61 distinct locations. The measurement device and the data are described in detail in [35]. The test subject was asked to blink and make horizontal eye saccades in order to produce typical ocular artifacts and bite the teeth for 20 seconds in order to create myographic artifacts. This 122-dimensional input data was first reduced to 20 dimensions by PCA, in order to eliminate noise and "bumbs", which appear in the data if the dimensionality is not sufficiently reduced [18]. Then a band-pass filtering was performed with the pass band between 0.5 and about 45 Hz. This eliminated most of the powerful low-frequency noise and the effect of the power grid at 50 Hz. The topographic ICA algorithm was then run on the data using a one dimensional ring-shaped topography. The neighbourhood was formed by convolving a vector of three ones with itself four times.

The resulting separated signals are shown in figure 3.6. The signals themselves are very similar to those found in [35]. As for the topographic organization, we can see that

1. The signals corresponding to bites (#9 - #15) are now adjacent. When computing the field patterns corresponding to these signals, one can also see that the signals are ordered according to whether they come from the left or the right side of the head.
2. Two signals corresponding to eye artifacts are also adjacent (#18 and #19). The signal # 18 corresponds to horizontal eye saccades and the signals #19 to eye-blinks.
3. A signal which seems to relate to eye activity has been separated into #17.

We can also see signals that do not seem to have any meaningful topographic relations, probably because they are quite independent from the rest of the signals. These include the heart beat (signal #7) and a signal corresponding to a digital watch which was at a distance of 1m from the magnetometer (signal #6). Thus topographic ICA orders the signals mainly into two clusters, one created by the signals coming from the muscle artifact and the other by eye muscle activity.

We see that topographic ICA finds largely the same components as those found by ICA in [35]. Using topographic ICA has the advantage, however, that signals are grouped together according to their dependence content. For example, topographic ICA suggested the interpretation of the signal #17 as related to eye activity, which was not clear from ICA.

3.7 Conclusion

We introduced topographic ICA, which is a generative model that combines topographic mapping with ICA. As in all topographic mappings, the distance in the representation space (on the topographic "grid") is related to the distance of the represented components. In topographic ICA, the distance between represented components is defined by the mutual information implied by the higher-order correlations, which gives the natural distance measure in the context of ICA. This is in contrast to practically all existing topographic mapping methods, where the distance is defined by basic geometrical relations like Euclidean distance or correlation, as in e.g. [23,4,11]. In fact, our principle makes it possible to define a topography even among a set of orthogonal vectors, whose Euclidean distances are all equal.

To estimate the model in practice, we considered a simple gradient learning rule of the likelihood.[3] This leads to an interesting form of Hebbian learning, where the Hebbian term is modulated by top-down feedback.

The utility of this novel model of topographic organization is clearly seen with natural image data, where topographic ICA gives a linear decomposition into Gabor-like linear features. In contrast to ordinary ICA, the higher-order

---

[3] In the generative model in section 3.3.3, this objective function is actually an approximation of the likelihood.

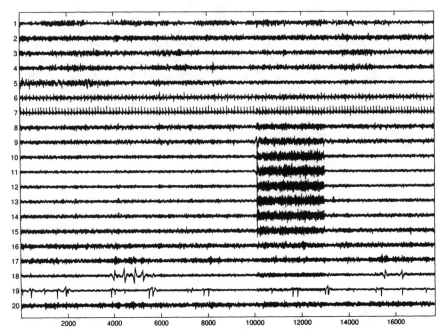

Fig. 3.6. The separated source signals found by topographic ICA from MEG data.

dependencies that linear ICA could not remove define a topographic order such that nearby cells tend to be active at the same time. This implies also that the neighbourhoods have properties similar to those of complex cells. Our model thus shows simultaneous emergence of complex cell properties and topographic organisation. These two properties emerge from the very same principle of defining topography by higher-order correlations. Similar results were obtained for speech feature extraction. In blind source separation, the method helps in interpreting the results with high-dimensional data and may reveal new aspects of the data.

References

1. S. I. Amari, A. Cichocki, and H. H. Yang. A new learning algorithm for blind signal separation. In *Advances in Neural Information Processing Systems 8*. 757–763. MIT Press, Cambridge, MA, 1996.
2. A. J. Bell and T. J. Sejnowski. An information-maximization approach to blind separation and blind deconvolution. *Neural Computation*, **7**:1129–1159, 1995.
3. A. J. Bell and T. J. Sejnowski. The independent components of natural scenes are edge filters. *Vision Research*, **37**:3327–3338, 1997.
4. C. M. Bishop, M. Svensen, and C. K. I. Williams. GTM: The generative topographic mapping. *Neural Computation*, **10**:215–234, 1998.

5. J.-F. Cardoso. Multidimensional independent component analysis. In *Proc. IEEE Int. Conf. on Acoustics, Speech and Signal Processing (ICASSP'98)*, Seattle, WA, 1998.
6. J. F. Cardoso. Entropic contrasts for source separation. In S. Haykin, editor, *Adaptive Unsupervised Learning*. 1999.
7. J. F. Cardoso and B. Hvam Laheld. Equivariant adaptive source separation. *IEEE Trans. on Signal Processing*, 44(12):3017–3030, 1996.
8. A. Cichocki and R. Unbehauen. Robust neural networks with on-line learning for blind identification and blind separation of sources. *IEEE Trans. on Circuits and Systems*, 43(11):894–906, 1996.
9. P. Comon. Independent component analysis – a new concept? *Signal Processing*, 36:287–314, 1994.
10. N. Delfosse and P. Loubaton. Adaptive blind separation of independent sources: a deflation approach. *Signal Processing*, 45:59–83, 1995.
11. G. J. Goodhill and T. J. Sejnowski. A unifying objective function for topographic mappings. *Neural Computation*, 9(6):1291–1303, 1997.
12. A. Hyvärinen. Fast and robust fixed-point algorithms for independent component analysis. *IEEE Trans. on Neural Networks*, 10(3):626–634, 1999.
13. A. Hyvärinen. Sparse code shrinkage: Denoising of nongaussian data by maximum likelihood estimation. *Neural Computation*, 11(7):1739–1768, 1999.
14. A. Hyvärinen. Survey on independent component analysis. *Neural Computing Surveys*, 2:94–128, 1999.
15. A. Hyvärinen and P. O. Hoyer. Topographic independent component analysis. 1999. Submitted, available at http://www.cis.hut.fi/~aapo/.
16. A. Hyvärinen and P. O. Hoyer. Emergence of phase and shift invariant features by decomposition of natural images into independent feature subspaces. *Neural Computation*, 2000. (In press).
17. A. Hyvärinen and E. Oja. A fast fixed-point algorithm for independent component analysis. *Neural Computation*, 9(7):1483–1492, 1997.
18. A. Hyvärinen, J. Särelä, and R. Vigário. Spikes and bumps: Artefacts generated by independent component analysis with insufficient sample size. In *Proc. Int. Workshop on Independent Component Analysis and Signal Separation (ICA'99)*, 425–429. Aussois, France, 1999.
19. C. Jutten and J. Herault. Blind separation of sources, part I: An adaptive algorithm based on neuromimetic architecture. *Signal Processing*, 24:1–10, 1991.
20. J. Karhunen, E. Oja, L. Wang, R. Vigário, and J. Joutsensalo. A class of neural networks for independent component analysis. *IEEE Trans. on Neural Networks*, 8(3):486–504, 1997.
21. K. Kiviluoto and E. Oja. S-Map: A network with a simple self-organization algorithm for generative topographic mappings. In *Advances in Neural Information Processing Systems*, 10. The MIT Press,Cambridge, MA, 1998.
22. T. Kohonen. Self-organized formation of topologically correct feature maps. *Biological Cybernetics*, 43(1):56–69, 1982.
23. T. Kohonen. *Self-Organizing Maps*. Springer-Verlag, New York, 1995.
24. T. Kohonen. Emergence of invariant-feature detectors in the adaptive-subspace self-organizing map. *Biological Cybernetics*, 75:281–291, 1996.
25. J. K. Lin. Factorizing multivariate function classes. In *Advances in Neural Information Processing Systems*, 10, 563–569. MIT Press, Cambridge, MA, 1998.

26. E. Oja. The nonlinear PCA learning rule in independent component analysis. *Neurocomputing*, **17**(1):25–46, 1997.
27. B. A. Olshausen and D. J. Field. Emergence of simple-cell receptive field properties by learning a sparse code for natural images. *Nature*, **381**:607–609, 1996.
28. B. A. Olshausen and D. J. Field. Sparse coding with an overcomplete basis set: A strategy employed by V1? *Vision Research*, **37**:3311–3325, 1997.
29. P. Pajunen. Blind source separation using algorithmic information theory. *Neurocomputing*, **22**:35–48, 1998.
30. D.-T. Pham, P. Garrat, and C. Jutten. Separation of a mixture of independent sources through a maximum likelihood approach. In *Proc. EUSIPCO*, 771–774, 1992.
31. D. Pollen and S. Ronner. Visual cortical neurons as localized spatial frequency filters. *IEEE Trans. on Systems, Man, and Cybernetics*, **13**:907–916, 1983.
32. E. P. Simoncelli and O. Schwartz. Modeling surround suppression in V1 neurons with a statistically-derived normalization model. In *Advances in Neural Information Processing Systems 11*, 153–159. MIT Press, Cambridge, MA, 1999.
33. N. V. Swindale. The development of topography in the visual cortex: a review of models. *Network*, **7**(2):161-247, 1996.
34. J. H. van Hateren and A. van der Schaaf. Independent component filters of natural images compared with simple cells in primary visual cortex. *Proc. Royal Society ser. B*, **265**:359–366, 1998.
35. R. Vigário, V. Jousmäki, M. Hämäläinen, R. Hari, and E. Oja. Independent component analysis for identification of artifacts in magnetoencephalographic recordings. In *Advances in Neural Information Processing Systems 10*, 229–235. MIT Press, Cambridge, MA, 1998.

4 The Independence Assumption: Dependent Component Analysis

Allan Kardec Barros

4.1 Introduction

Redundancy reduction as a form of neural coding has been a topic of great research interest since the early sixties. A number of strategies have been proposed, but the one which is attracting most attention recently assumes that this coding is carried out so that the output signals are as independent as possible. In this work, we go one step further and propose an algorithm to separate non-orthogonal signals (i.e., dependent signals) based on the minimization of the output mutual spectral overlap. Indeed, separating independent sources turns to be a special case of this strategy. Moreover, we show that this principle can also be used to separate spectrally overlapping signals by exploiting their higher-order cyclostationary properties. We also suggest a numerically-efficient algorithm which searches for the learning step size in a way that avoids divergence.

There is much research interest recently in algorithms which exploit the principle of redundancy reduction as a form of coding strategy in neurons. For example, Barlow [4,5] suggested that the neurons could code their outputs so that they are as statistically independent as possible. Indeed, this reasoning was used by many researchers in the field of *blind source separation* (BSS), particularly in the recent approach called independent component analysis (ICA). Given that a linear mixture of source signals is observed, ICA attempts to extract those sources assuming that they are *mutually independent*.[1] ICA has been applied to a number of areas, including speech processing, telecommunications, chemistry, and biomedical signal processing among others.

In this work, we go in similar direction. However, we extend the framework to include also *non-orthogonal* signals. In other words, we drop the assumption of statistical independence among the sources, although we suggest a strategy for decreasing redundancy, in the sense that the output signals should have minimal mutual spectral overlap. Therefore, in order to differentiate it from ICA, we call this type of approach *Dependent Component Analysis* (DCA).

This work may be understood as a preliminary step towards codes that include non-orthogonal signals, which is clearly important in some applications such as those found in the works of Nuzillard and Nuzillard[17], where

---

[1] See references for further information about ICA.

the authors had to use the available *a priori* information in order to separate real world non-orthogonal signals.

Given this, we propose here a learning rule to recover the original non-orthogonal signals from a linear mixture. Indeed, we will see that the task of extracting statistically independent components can be regarded as a special case of the proposed algorithm. The algorithm is based on spectral coherence minimization of two signals.

Instead of minimizing the cost function using the steepest-descent algorithms, we propose a numerically-efficient rule for searching for the minimum of a cost function in a three-dimensional space.

4.2   Blind Source Separation by DCA

Blind source separation may be understood as follows. Consider n source signals $s = [s_1, s_2, .., s_n]^T$ arriving at n receivers. Each receiver gets a linear combination of the signals, so that we have

$$x = As \tag{4.1}$$

where A is an $n \times n$ invertible matrix.

The purpose of BSS is to find a linear combination of the elements of x which gives as output the elements of s, possibly re-scaled and permuted. This output, $u = Wx$ is found using a weight matrix W, so that $WA = DP$, where D is a diagonal and P is a permutation matrix.

To find matrix W, we suggest to minimising the averaged coherence function (ACF) defined as

$$\zeta_u = \frac{1}{2\pi} \int_{-\pi}^{\pi} \frac{|S_{u12}(\omega)|^2}{S_{u11}(\omega) S_{u22}(\omega)} d\omega \tag{4.2}$$

where $S_{u12}(\omega)$ is the cross-power spectrum between u_1 and u_2, and $S_{uii}(\omega)$ is the power spectrum of u_i, and ω is the frequency. It is important to notice three properties of the ACF:

1. $0 < \zeta_u < 1$. This follows directly from the spectral inequality $|S_{u12}(\omega)|^2 \leq S_{u11}(\omega) S_{u22}(\omega)$.
2. If the elements of s are mutually independent, $\zeta_s = 0$. The proof follows directly from the fact that if the elements of s are independent, they are therefore orthogonal, thus $S_{s12}(\omega) = 0$, $\forall \omega$.
3. If $\zeta_u \geq \zeta_s$, for any invertible W, then the minimum $\zeta_u = \zeta_s$ occurs when $WA = DP$. The steps for this proof can be found in the appendix. Notice that while for non-orthogonal signals this relation is case-dependent, for statistically independent sources this is always true.

Defining $\mathbf{w}_i$ as the i-th row of $\mathbf{W}$, the elements of $\mathbf{u}$ can be written as $u_i = \mathbf{w}_i^T \mathbf{x}$. Remembering that the power spectrum $S_{uii}(\omega)$ is a function of the autocorrelation $r_{uii}(\tau) = E[u_{ii}(t+\tau)u_{ii}(t)]$ by the following equation [19]

$$S_{uii}(\omega) = \int_{-\infty}^{\infty} r_{uii}(\tau)e^{-j\omega\tau}d\tau \tag{4.3}$$

we find the relation

$$S_{uii}(\omega) = \int_{-\infty}^{\infty} \mathbf{w}_i^T \mathbf{R}_\mathbf{x}(\tau)e^{-j\omega\tau}\mathbf{w}_i d\tau \tag{4.4}$$

$$\mathbf{R}_\mathbf{x}(\tau) = \begin{pmatrix} r_{x11}(\tau) & r_{x12}(\tau) \\ r_{x21}(\tau) & r_{x22}(\tau) \end{pmatrix}$$

From equiation 4.3, we can easily find the following equation,

$$S_{uii}(\omega) = \mathbf{w}_i^T \left\{ \int_{-\infty}^{\infty} \mathbf{R}_\mathbf{x}(\tau)e^{-j\omega\tau}d\tau \right\} \mathbf{w}_i = \mathbf{w}_i^T \mathbf{S}_\mathbf{x}(\omega)\mathbf{w}_i \tag{4.5}$$

Thus, we have the ACF for the output given by

$$\zeta_u = \frac{1}{2\pi} \int_{-\pi}^{\pi} \frac{\left| \mathbf{w}_1^T \mathbf{S}_\mathbf{x}(\omega)\mathbf{w}_2 \right|^2}{\mathbf{w}_1^T \mathbf{S}_\mathbf{x}(\omega)\mathbf{w}_1 \mathbf{w}_2^T \mathbf{S}_\mathbf{x}(\omega)\mathbf{w}_2} d\omega \tag{4.6}$$

The gradient of this function is thus

$$\frac{\partial \zeta_u}{\partial \mathbf{w}_i} = \int_{-\pi}^{\pi} \frac{2\mathbf{S}\mathbf{w}_j(\mathbf{w}_i^T \mathbf{S}\mathbf{w}_j)}{(\mathbf{w}_i^T \mathbf{S}\mathbf{w}_i)(\mathbf{w}_j^T \mathbf{S}\mathbf{w}_j)} - \int_{-\pi}^{\pi} \frac{(\mathbf{w}_i^T \mathbf{S}\mathbf{w}_j)^2 (\mathbf{S} + \mathbf{S}^T)\mathbf{w}_i}{(\mathbf{w}_i^T \mathbf{S}\mathbf{w}_i)(\mathbf{w}_j^T \mathbf{S}\mathbf{w}_j)} \tag{4.7}$$

which can be minimised using a steepest-descent (SD) algorithm, for example. However, the choice of step size for the SD algorithm is a hard task: a sufficiently small value may lead to the solution, but at a slower convergence rate; and the calculation of an upper bound can be particularly difficult because of the non-linear cost function. Thus we suggest in the next section an efficient numerical algorithm for minimisation of cost functions such as equation 4.2 or equation 4.7.

4.3 The "Cyclone" Algorithm

A cyclone is "a system of winds rotating inwards to an area of low barometric pressure"[18]. We use this idea to search for the minimum of ACF, as given by equation 4.2 in a three-dimensional space.

First, we define the cost function $\zeta_u(\theta_1^k, \theta_2^k)$ as dependent on two variables θ_1^k and θ_2^k, where k is the iteration number for discrete learning (see figure 4.1). Secondly, we define a spiral with a *spiral point* given by $\Theta_* = [\theta_1^* \ \theta_2^*]^T$, then calculate $\zeta_u(\theta_1^k, \theta_2^k)$ for $k = 1, 2, \ldots$, until a smaller value $\zeta_u(\theta_1^i, \theta_2^i)$, is reached. The learning continues by moving the spiral point to the new vector, i.e., $\Theta_* = \Theta_i$.

Thus, the algorithm can be simply written as

$$\Theta_k = \Theta_* + \mu_k \Delta d\Theta \tag{4.8}$$

where μ and Δ are given by

$$\mu_k = \beta \mu_{k-1}, \qquad\qquad 0 < \beta < 1$$
$$\Delta = \begin{pmatrix} \cos(\alpha\pi) & -\sin(\alpha\pi) \\ \sin(\alpha\pi) & \cos(\alpha\pi) \end{pmatrix} \quad 0 < \alpha < \frac{1}{2} \tag{4.9}$$

where β defines how fast to approach the center and α is the angle between the current and next points of the spiral.

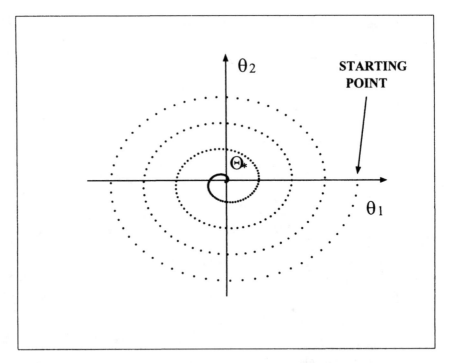

Fig. 4.1. Geometric interpretation for the algorithm. The learning moves from outside to inside, with k increasing as the learning approaches the spiral point Θ_*.

Thus, to find ACF, we suggest carrying out the following steps:

1. Choose $d\Theta = [d\theta_1 \; d\theta_2]^T$. μ_{min} and μ_{max} are the minimum and maximum bounds for μ, respectively. Set $k = 1$.

2. Do:

$$\mathbf{W}_k = \begin{pmatrix} 1 & \theta_1^k \\ \theta_2^k & 1 \end{pmatrix}$$

3. Calculate $\zeta_u(k)$ from equation 4.2 and $\mathbf{u} = \mathbf{W}\mathbf{x}$. If $\zeta_u(k) < \zeta_u(k-1)$, do $\Theta_* = \Theta_k$.
4. Update k. Update the weights as given by equation 4.8.
5. If $\mu_k < \mu_{min}$, do $\mu_k = \mu_{max}$.
6. Repeat steps 2 to 5 until the Euclidean distance between two consecutive Θ_* is smaller than a pre-defined error, after one complete turn around the spiral.

For small enough α, β and μ_{min}, the spiral approaches the area of a circle with radius μ_{max}. Convergence is assured as long as the condition

$$\frac{\partial^2 \zeta_u(\theta_1^k, \theta_2^k)}{\partial \theta_1^k \partial \theta_2^k} \neq 0, \ \forall k$$

is fulfilled. This condition implies that inside this area there is a $\zeta_u(\theta_1^k, \theta_2^k) < \zeta_u(\theta_1^*, \theta_2^*)$ and the algorithm will converge to a minimum, or that $\zeta_u(\theta_1^k, \theta_2^k) > \zeta_u(\theta_1^*, \theta_2^*)$ everywhere and $\zeta_u(\theta_1^*, \theta_2^*)$ is already a minimum. Then we can say that the algorithm is locally convergent.

4.4 Experimental Results

We have carried out extensively the classical simulation of randomly mixing source signals and extracting them by the proposed algorithm. Two signals taken from the MIT-BIH noise stress test database were mixed . They were composed of an electrocardiographic signal and a band-pass filtered version of the same signal, added to a randomly distributed Gaussian signal.

We set the parameters of the "cyclone" algorithm for the simulations above as $\beta = 1/2$, $\alpha = 1/4$ and $\mu_{max} = 1$. Figure 4.2 shows the movement of the *spiral point* on the surface of the averaged coherence function. Generally, the cyclone algorithm converged in less than 40 iterations.

As an efficiency factor, we can define $\mathbf{C} = \mathbf{W}\mathbf{A}$, which, in the case of a successful separation should be $\mathbf{C} = \mathbf{D}\mathbf{P}$, as we have seen above. Below we show one result of a simulation, for the DCA algorithm proposed here compared to the ICA algorithm proposed in [13]

ICA	DCA
$\begin{pmatrix} 0.00 & 1.00 \\ 1.00 & 0.31 \end{pmatrix}$	$\begin{pmatrix} 1.00 & 0.02 \\ 0.00 & 1.00 \end{pmatrix}$

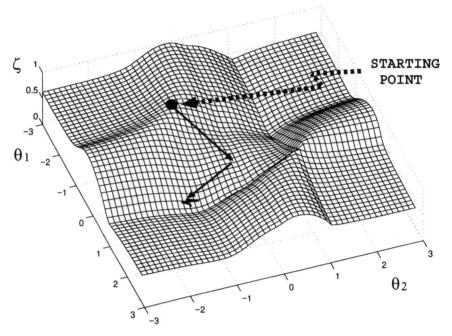

Fig. 4.2. The learning behaviour of the cyclone algorithm on the surface of the ACF, which is a function of two variables θ_1 and θ_2.

4.5   Higher-Order Cyclostationary Signal Separation

ACF property **3** may not always be satisfied, but we can still use non-linearities in some cases to separate spectrally overlapping signals. Here we show an example of two periodic signals which can be separated by using a function $\zeta_{f(u)}$, where $f(u)$ is a non-linear function of u.

This transformation can be regarded as a kind of exploitation of the higher-order cyclostationary (HOC) properties of the signals [12]. Indeed, we mixed one square with a sine wave, with the same fundamental frequency. By using $f(u) = u^2$, the separation was successfully obtained, as shown in figure 4.3.

4.6   Conclusion

The direction of this work is similar to that of Barlow [4,5]: trying to understand the learning strategy carried out by humans in the brain. Our assumption is that this strategy is carried out by redundancy reduction. However, instead of assuming that the source signals are mutually independent as is necessary for ICA, we go one step further and code the information in terms

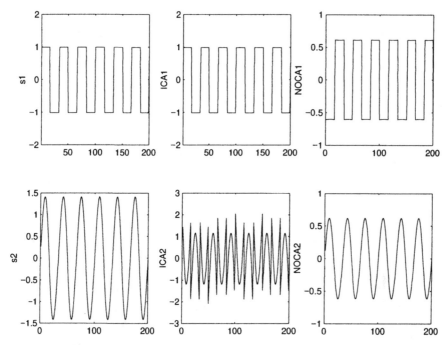

Fig. 4.3. Two source signals (left column) with the same fundamental frequency, the separation given by an ICA algorithm as in [13] (middle column) and the separation given by exploiting the HOC properties of the signals using DCA (right column).

of its *spectral* redundancy, which also includes independent signals. Moreover, it is known that, for example, the hearing system codes the received information in terms of *frequency*, i.e., the inner ear can be understood as a number of band-pass filters through which the incoming signal is processed. Thus, it is not difficult to imagine that the coding strategy proposed here could be another form which neurons use to process information by redundancy reduction.

We saw in the simulations that the algorithm works efficiently under rather mild assumptions compared to independent component analysis. However, we believe that further progress can be carried out by using assumptions stronger than that of ACF property **3**. Moreover, it is necessary to generalize the cost function in order to include mixtures of more than two source signals.

Another contribution of this work is the cyclone algorithm, which is numerically efficient at minimising a cost function which depends on two variables. One important aspect of this algorithm is that it does not diverge, as it can occur in a number of steep-descent algorithms. Indeed, this was one of our intentions when we proposed it: ingore the step size and its bounds. However, one should notice that just as we used the cyclone algorithm to find the minimum of ACF, we could equally have used it to find the minimum or maximum of its gradient.

4.7   Appendix: Proof of ACF Property 3

First, let us define matrix $\mathbf{C} = \mathbf{WA}$. Thus, following the steps from equation 4.3 to equation 4.6, we can find the ACF for the output as given by

$$\zeta_u = \frac{1}{2\pi} \int_{-\pi}^{\pi} \frac{\left|\mathbf{c}_1^T \mathbf{S_s}(\omega) \mathbf{c}_2\right|^2}{\mathbf{c}_1^T \mathbf{S_s}(\omega) \mathbf{c}_1 \mathbf{c}_2^T \mathbf{S_s}(\omega) \mathbf{c}_2} d\omega, \tag{4.10}$$

where $\mathbf{c}_i$ is the i-th row of $\mathbf{C}$ and

$$\mathbf{S_s}(\omega) = \begin{pmatrix} S_{s11}(\omega) & S_{s12}(\omega) \\ S_{s21}(\omega) & S_{s22}(\omega) \end{pmatrix}. \tag{4.11}$$

Developing equation 4.10 further using equation 4.11, one finds easily that $\zeta_u = \zeta_s$ iff $\mathbf{C} = \mathbf{DP}$. Moreover, one can easily find that this property is always valid for independent signals.

References

1. S. Amari, A. Cichocki, H. H. Yang. A new learning algorithm for blind signal separation. *In D. S. Touretzky, M. C. Mozer and M. E. Hasselmo, editors, Advances in Neural Information Processing Systems*, 8, MIT Press, Cambridge, MA, 1996.
2. S. Amari and A. Cichocki. Adaptive blind signal processing : neural network approaches, *Proceedings IEEE* (invited paper), **86**: 2026-2048, 1998.
3. Y. Baram and Z. Roth. Density shaping by neural networks with application to classification, estimation and forecasting. *CIS report no. 9420*. Center for Intelligent Systems, Technion, Israel, 1994.
4. H. B. Barlow, Possible principles underlying the transformations of sensory messages. In W. Rosenblith editor *Sensory Communication*. pp. 217-234. MIT Press, Cambridge, MA, 1961.
5. H. B. Barlow, Unsupervised learning. *Neural Computation*, **1**: 295 - 311, 1989.
6. A. K. Barros, A. Mansour and N. Ohnishi, Removing artifacts from ECG signals using independent components analysis". *Neurocomputing*, **22**: 173 - 186, 1998.
7. A. J. Bell and T. J. Sejnowski, An information-maximization approach to blind separation and blind deconvolution. *Neural Computation*, **7**: 1129 - 1159, 1995.
8. J-F. Cardoso and B. Hvam Laheld. Equivariant adaptive source separation.*IEEE Trans. on Signal Process*, 3017 - 3030, 1996.
9. A. Cichocki and L. Moszczynski. A new learning algorithm for blind separation of sources. *Electronics Letters*, **28**(21): 1986-1987, 1992.

10. P. Comon. Independent component analysis, a new concept? *Signal Processing*, **24**: 287 - 314, 1994.
11. G. Deco and W. Brauer. Nonlinear higher-order statistical decorrelation by volume-conserving neural architectures. *Neural Networks*, **8**: 525 - 535, 1995.
12. W. Gardner (editor). *Cyclostationarity in communications and signal processign*. IEEE Press, 1994.
13. A. Hyvärinen and E. Oja. A fast fixed-point algorithm for independent component analysis. *Neural Computation* **9**: 1483 - 1492, 1997.
14. C. Jutten and J. Hérault. Independent component analysis versus PCA. *Proc. EUSIPCO*, pp. 643 - 646, 1988.
15. S. Makeig, T-P. Jung, D. Ghahremani, A. J. Bell, T. J. Sejnowski. Blind separation of event-related brain responses into independent components. Proc. Natl. Acad. Sci. USA, **94**: 10979-10984, 1997.
16. J-P. Nadal and N. Parga. Non-linear neurons in the low noise limit: a factorial code maximises information transfer. *Network*, **5**: 565 - 581, 1994.
17. D. Nuzillard and J-M. Nuzillard. Blind source separation applied to nonorthogonal signals. *Proc. ICA '99*, 25 - 30, 1999.
18. J.A. Simpson, and E.S.C Weiner. *The Oxford English Dictionary*, 2nd edn. Clarendon Press, Oxford, 1989.
19. A. Papoulis. *Probability, random variables, and stochastic processes*. McGraw-Hill, 1991.

Part III

Ensemble Learning and Applications

5   Ensemble Learning

Harri Lappalainen and James W. Miskin

5.1   Introduction

This chapter gives a tutorial introduction to ensemble learning, a recently developed Bayesian method. For many problems it is intractable to perform inferences using the true posterior density over the unknown variables. Ensemble Learning allows the true posterior to be approximated by a simpler approximate distribution for which the required inferences are tractable. When we say we are making a model of a system, we are setting up a tool which can be used to make inferences, predictions and decisions. Each model can be seen as a hypothesis, or explanation, which makes assertions about the quantities which are directly observable and those which can only be inferred from their effect on observable quantities.

In the Bayesian framework, knowledge is contained in the conditional probability distributions of the models. We can use Bayes' theorem to evaluate the conditional probability distributions for the unknown parameters, y, given the set of observed quantities, x, using

$$p(y|x) = \frac{p(x|y)\,p(y)}{p(x)} \tag{5.1}$$

The prior distribution $p(y)$ contains our knowledge of the unknown variables before we make any observations. The posterior distribution $p(y|x)$ contains our knowledge of the system after we have made our observations. The likelihood, $p(x|y)$, is the probability that the observed data will be observed given a specific set of values for the unknown parameters.

There is no clear cut difference between the prior and posterior distributions, since after a set of observations the posterior distribution becomes the prior for another set of observations.

In order to make inferences based on our knowledge of the system, we need to marginalise our posterior distribution with respect to the unknown parameters of the model. For instance in order to obtain the average values of the unknown parameters we would need to perform the expectation

$$\bar{y} = \int y p(y|x)\,dy \tag{5.2}$$

Alternatively we may be trying to use our model of the system to make decisions about which action to take. In this case we would like to choose the action which maximises some utility function and the expected utility is found by marginalising the utility function over the posterior density of the models. An example would be hypothesis testing, where we have a number of explanations for the cause of a set of observed data. By having a different model for each hypothesis, we could choose the model that maximises the expected utility.

In this chapter we shall propose the idea of considering the posterior distribution to be the result of any experiment instead of just considering a point in the model space. Section 5.3 will discuss methods of approximating the posterior when it is intractable to make inferences based on the true posterior. Section 5.4 will introduce the idea of using ensemble learning to approximate the true posterior by a simpler separable distribution. Section 5.5 will discuss the construction of probabilistic models, both in supervised and unsupervised learning. Section 5.6 will give examples of using ensemble learning in both a fixed form and free form approximation.

5.2   Posterior Averages in Action

Probability theory tells us that the optimal generalisation is the one resulting from a Bayesian approach. Overfitting to the data means that we are making conclusions that the data does not support. Alternatively underfitting means that our conclusions are too diffuse.

Overfitting is an artifact resulting from choosing only one explanation (model) for the observations. Figure 5.1 shows a hypothetical posterior distribution. If the model is chosen to maximise the posterior probability then the model will be chosen to be in the narrow peak. The problem is that the peak only contains a fraction of the total probability mass. This means that the model will explain the observations very well, but will be very sensitive to the values of the parameters and so may not explain further observations. When making predictions and decisions it is the position of the probability mass that counts and not the position of a maximum.

In order to solve these problems it is necessary to average over all possible hypotheses. That is we should average over all possible models, but weight them by the posterior distribution that we obtain from our observations.

It is important to note that averaging does not mean computing the average of parameter values and then using that as the best guess. For instance in digit recognition the posterior distribution may have a peak in the model for 1s and for 9s. Averaging over the posterior does not mean that you compute the average digit as a 5 and then use that as the best guess, it means that you should prepare for the possibility that the digit can be either 1 or 9 .

Figure 5.2 shows a schematic example of fitting a Multi-Layer Perceptron (MLP) to a set of data. The data are fairly smooth and have only a little

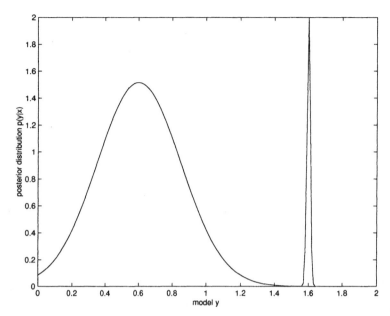

Fig. 5.1. Schematic illustration of overfitting to the data. If the true posterior over the models, y, is of this form then choosing the model that maximises the posterior probability will mean choosing the model in the sharp peak. But near the maximum the posterior density is sharp and so highly dependent on the model parameters. Therefore the model will explain the observed data very well, but may not generalise to further observations.

noise, but there is a region of missing data. The best regularised MLP is able to fit the data quite well in those areas where there are data points and so the noise level will be estimated to be low. Therefore the MLP will give tight error bars even in the region where there is no data to support the fit. This is overfitting because the conclusion is more specific than the data supports. If we were to obtain some data in the missing region later, it is plausible that it would lie significantly outside the MLP fit and so the posterior probability of the chosen model could rapidly drop as more data is obtained.

Instead of choosing a specific MLP, it is better to average over several MLPs. In this case we would find that there are multiple explanations for the missing data. Therefore the average fit will give tight error bars where there is data and broad error bars in the regions where there is no data.

If the estimate for the level of noise in the data were estimated to be too high then the error bars for the areas with missing data might be good but the error bars for the areas with data would be too broad and we would suffer from underfitting. Alternatively if the noise level is estimated to be too low, the error bars will be too narrow where there is data. For intermediate

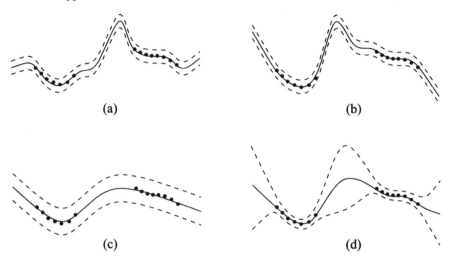

Fig. 5.2. A schematic illustration of averaging over the posterior of models. The data points are denoted by dots and the black lines show the MLP fit. The dashed lines show the error bars the model gives. The models in plots a and b assume fairly small noise. The model in plot c assumes larger noise. The average of the models over the posterior pdf of the models gives larger error bars in the areas where there are no observations.

noise levels we could suffer from overfitting in some regions and underfitting in others.

In order to fix this we could allow the noise level to be part of our model and so the noise level could vary. The problems with overfitting would now be changed to overfitting the noise model. The solution is to average the noise under the posterior distribution which would include the MLP parameters and the noise.

5.3   Approximations of Posterior PDF

As we have shown earlier, the posterior distribution is a synonym for our available knowledge of a system after we have made a set of observations. In order to use this knowledge we will typically be required to marginalise the posterior or to evaluate expectations of functions under the posterior distribution. In many problems it is intractable to perform the necessary integrals.

If we look again at Bayes equation we have

$$p(y|x) = \frac{p(x|y)p(y)}{p(x)} \tag{5.3}$$

It is easy to compute one point in the joint density $p(x, y)$ (the numerator of the posterior distribution) but in general evaluating the denominator, $p(x)$, is difficult. Similarly marginalising the posterior distribution is difficult.

Therefore it is necessary to approximate the posterior density by a more tractable form for which it is possible to perform any necessary integrals. We cannot take a point estimate (such as the MAP estimate) because this leads to overfitting as shown earlier. This is because the MAP estimate does not guarantee a high probability mass in the peak of the posterior distribution and so the posterior distribution may be sharp around the MAP estimate. We would like to approximate the probability mass of the posterior.

There are in general two types of approximation that retain the probability mass of the true posterior distribution: the stochastic approximation and the parametric approximation. In a stochastic approximation (such as Markov-chain Monte Carlo method) the aim is to perform the integrations by drawing samples from the true posterior distribution [3]. The average of any function is then found by finding the average value of the function given all of the samples from the posterior.

In a parametric approximation (such as the Laplace approximation [6]) the posterior distribution is approximated by an alternative function (such as a Gaussian) such that it is much simpler to perform any necessary approximations.

The problem with the stochastic methods is that, when performing a stochastic approximation, it is necessary to wait until the sampler has sampled from all of the mass of the posterior distribution. Therefore testing for convergence can be a problem. The problem with the parametric approach is that the integrals being performed are not exactly the same as those that would be performed when the true posterior is used. Therefore while the stochastic approximation has to give the correct answer (eventually) the parametric approxmation will give an approximate answer soon (it is the 'quick and dirty' method).

Model selection can be seen as a special form of approximating the posterior distribution. The posterior distribution could contain many peaks, but when there is lots of data, most of the probability mass is typically contained in a few peaks of the posterior distribution. Model selection means using only the most massive peaks and discarding the remaining models.

5.4 Ensemble Learning

Ensemble learning [1] is a recently introduced method for parametric approximation of the posterior distributions where Kullback-Leibler information [2,5], is used to measure the misfit between the actual posterior distribution and its approximation. Let us denote the observed variables by x and the unknown variables by y. The true posterior distribution $p(y|x)$ is approximated with the distribution $q(y|x)$ by minimising the Kullback-Leibler information.

$$D(q(y|x)\|p(y|x)) = \int q(y|x) \ln \frac{q(y|x)}{p(y|x)} dy$$

$$= \int q(y|x) \ln \frac{p(x)q(y|x)}{p(x,y)} dy$$

$$= \int q(y|x) \ln \frac{q(y|x)}{p(x,y)} dy + \ln p(x) \tag{5.4}$$

The Kullback-Leibler information is greater than or equal to zero, with equality if and only if the two distributions, $p(y|x)$ and $q(y|x)$, are equivalent. Therefore the Kullback-Leibler information acts as a distance measure between the two distributions.

If we note that the term $p(x)$ is a constant over all the models, we can define a cost function $C_y(x)$ which we must minimise to obtain the optimum approximating distribution

$$C_y(x) = D(q(y|x)\|p(y|x)) - \ln p(x) = \int q(y|x) \ln \frac{q(y|x)}{p(x,y)} dy \tag{5.5}$$

We shall adopt the notation that the subindex of C denotes the variables that are marginalised over the cost function. In general, they are the unknown variables of the model. The notation also makes explicit that the cost function $C_y(x)$ gives an upper bound for $-\ln p(x)$. Here we use the same notation as with probability distributions, that is $C_y(x|z)$ means

$$C_y(x|z) = D(q(y|x,z)\|p(y|x,z)) - \ln p(x|z)$$

$$= \int q(y|x,z) \ln \frac{q(y|x,z)}{p(x,y|z)} dy \tag{5.6}$$

and thus yields the upper bound for $-\ln p(x|z)$.

Ensemble learning is practical if the terms $p(x,y)$ and $q(y|x)$ of the cost function $C_y(x)$ can be factorised into simple terms. If this is the case, the logarithms in the cost function split into sums of many simple terms. By virtue of the definition of the models, the likelihood and priors are both likely to be products of simpler distributions; $p(x,y)$ typically factorises into simple terms. In order to simplify the approximating ensemble, q is also modelled as a product of simple terms.

The Kullback-Leibler information is a global measure, providing that the approximating distribution is a global distribution. Therefore the measure will be sensitive to probability mass in the true posterior distribution rather than the absolute value of the distribution itself.

Training the approximating ensemble can be done by assuming a fixed parametric form for the ensemble (for instance assuming a product of Gaussians). The parameters of the distributions can then be set to minimise the cost function.

An alternative method is to assume a separable form for the approximating ensemble. The distributions themselves can then be found by performing a functional minimisation of the cost function with respect to each distribution in the ensemble. While this method must always give ensembles with equivalent or lower misfits than those obtained by assuming a parametric form, the distributions that are obtained are not always tractable and so the fixed form method may be more useful.

5.4.1   Model Selection in Ensemble Learning

Recall that the cost function $C_y(x|H)$ can be translated into the lower bound for $p(x|H)$. Since $p(H|x) = p(x|H)p(H)/p(x)$, it is natural that $C_y(x|H)$ can be used for model selection also by equating

$$p(H|x) \approx \frac{e^{-C_y(x|H)}P(H)}{\sum_{H'} e^{-C_y(x|H')}P(H')} \tag{5.7}$$

In fact, we can show that the above equation gives the best approximation for $p(H|x)$ in terms of $C_{y,H}(x)$, the Kullback-Leibler divergence between $q(y, H|x)$ and $p(y, H|x)$, which means that the model selection can be done using the same principle of approximating the posterior distribution as learning parameters.

Without losing any generality from $q(y, H|x)$, we can write

$$q(y, H|x) = Q(H|x)q(y|x, H) \tag{5.8}$$

Now the cost function can be written as

$$
\begin{aligned}
C_{y,H}(x) &= \sum_H \int q(y, H|x) \ln \frac{q(y, H|x)}{p(x, y, H)} dy \\
&= \sum_H Q(H|x) \int q(y|x, H) \ln \frac{Q(H|x)q(y|x, H)}{P(H)p(x, y|H)} dy \\
&= \sum_H Q(H|x) \left[\ln \frac{Q(H|x)}{P(H)} + C_y(x|H) \right]
\end{aligned} \tag{5.9}
$$

Minimising $C_{y,H}(x)$ with respect to $Q(H|x)$ under the constraint

$$\sum_H Q(H|x) = 1 \tag{5.10}$$

yields

$$Q(H|x) = \frac{e^{-C_y(x|H)}P(H)}{\sum_{H'} e^{-C_y(x|H')}P(H')} \tag{5.11}$$

Substituting this into equation 5.9 yields the minimum value for $C_{y,H}(x)$ which is

$$C_{y,H}(x) = -\ln \sum_H e^{-C_y(x|H)} P(H) \qquad (5.12)$$

If we wish to use only a part of different model structures H, we can try to find those H which would minimise $C_{y,H}(x)$. It is easy to see that this is accomplished by choosing the models corresponding to $C_y(x|H)$. A special case is to use only one H corresponding to the smallest $C_y(x|H)$.

5.4.2   Connection to Coding

The cost function can be derived within MDL framework. The intuitive idea behind MDL is that in order to be able to encode data compactly one has to find a good model for it. Coding is not actually done; only the formula for the code length is needed. The optimal code length of x is $L(x) = -\log P(x)$ bits, and therefore coding has a close connection to the probabilistic framework. Some people prefer to think in terms of coding lengths, some in term of probabilities.

In coding, the sender and the receiver have to agree on the structure of the message in advance. This corresponds to having to state one's priors in the Bayesian framework.

A simple example: let x be the observation to be coded and y be a real valued parameter. We first encode y with accuracy dy. This part of the message takes $L(y) = -\log p(y)dy$ bits. Then we encode x with accuracy dx using the value y. This takes $L(x|y) = -\log p(x|y)dx$ bits. The accuracy of the encoding of the observations can be agreed on in advance, but there is a question of how to determine the accuracy dy. The second part of the code would be shortest if y is in exactly the maximum likelihood solution. If y is encoded in too high accuracy, however, the first part of the code will be very long. If y is encoded in too low accuracy, the first part will be short but deviations from the optimal y will increase the second part of the code.

The bits-back argument [1] overcomes the problem by using infinitesimally small dy but picking the y from a distribution $q(y)$ and encoding a secondary message in the choice of y. Since the distribution $q(y)$ is not needed for decoding x, both the sender and the receiver can run the same algorithm for determining $q(y)$ from x. After finding out $q(y)$, the receiver can decode the secondary message which can have $-\log q(y)dy$ bits. As these bits will be returned in the end, the expected message length is

$$
\begin{aligned}
L(x) &= E_q\{-\log p(x|y)dx - \log p(y)dy + \log q(y)dy\} \\
&= D(q(y)\|p(y|x)) - \log p(x)dx = C_y(x) - \log dx \qquad (5.13)
\end{aligned}
$$

The number of bits used for individual parameter θ can be measured by looking at $D(q(\theta)\|p(\theta))$.

It is interesting to notice that although the message has two parts, y and x, if $q(y)$ is chosen to be $p(y|x)$, the expected code length is equal to that of the optimal one-part message where the sender encodes x directly using the marginal probability $p(x) = \int p(x|y)p(y)dy$.

5.4.3   EM and MAP

The Expectation-Maximisation (EM) algorithm can be seen as a special case of ensemble learning. The set-up in EM is the following: Suppose we have a probability model $p(x, y|\theta)$. We observe x but y remains hidden. We would like to estimate θ with maximum likelihood, i.e., maximise $p(x|\theta)$ w.r.t. θ, but suppose the structure of the model is such that integration over $p(x, y|\theta)$ is difficult, i.e., it is difficult to evaluate $p(x|\theta) = \int p(x, y|\theta)dy$.

We take the cost function $C_y(x|\theta)$ and minimise it alternately with respect to θ and $q(y|x, \theta)$. The ordinary EM algorithm will result when $q(y|x, \theta)$ has a free form in which case $q(y|x, \theta)$ will be updated to be $p(y|x, \hat{\theta})$, where $\hat{\theta}$ is the current estimate of θ. The method is useful if integration over $\ln p(x, y|\theta)$ is easy, which is often the case. This interpretation of EM was given by [4].

EM algorithm can suffer from overfitting because only point estimates for the parameters θ are used. Even worse is to use a *maximum a posterior* (MAP) estimator where one finds the θ and y which maximise $p(y, \theta|x)$. Unlike maximum likelihood estimation, MAP estimation is not invariant under reparameterisations of the model. This is because MAP estimation is sensitive to probability density which changes nonuniformly if the parameter space is changed non-linearly.[1]

MAP estimation can be interpreted in the ensemble learning framework as minimising $C_{y,\theta}(x)$ and using delta distribution as $q(y, \theta|x)$. This makes the integral $\int q(y, \theta|x) \ln q(y, \theta|x)dyd\theta$ infinite. It can be neglected when estimating θ and y because it is constant with respect to $\hat{y}$ and $\hat{\theta}$, but the infinity of the cost function shows that delta distribution, i.e. a point estimator, is a bad approximation for a posterior density.

5.5   Construction of Probabilistic Models

In order to apply the Bayesian approach to modelling, the model needs to be given in probabilistic terms, which means stating the joint distribution of all the variables in the model. In principle, any joint distribution can be regarded as a model, but in practice, the joint distribution will have a simple form.

---

[1] MAP estimation can be made invariant under reparameterisations by fixing the parameterisation. A natural way to do this is to use a parameterisation which makes the Fisher information matrix of the model constant. Then it is probable that the peaks in the posterior have approximately similar widths and are more or less symmetrical.

As an example, we shall see how a generative model turns into a probabilistic model. Suppose we have a model which tells how a sequence $y = y(1), \ldots, y(t)$ transforms into sequence $x = x(1), \ldots, x(t)$.

$$x(t) = f(y(t), \theta) + n(t) \tag{5.14}$$

This is called a generative model for x because it tells explicitly how the sequence x is generated from the sequence y through a mapping f parameterised by θ. As it is usually unrealistic to assume that it would be possible to model all the things affecting x exactly, the models typically include a noise term $n(t)$.

If $y(t)$ and θ are given, then $x(t)$ has the same distribution as $n(t)$ except that it is offset by $f(y(t), \theta)$. This means that if $n(t)$ is Gaussian noise with variance σ^2, equation 5.14 translates into

$$x(t) \sim N(f(y(t)), \sigma^2) \tag{5.15}$$

which is equivalent to

$$p(x(t)|y(t), \theta, \sigma) = \frac{1}{\sqrt{2\pi\sigma^2}} e^{-\frac{[x(t)-f(y(t),\theta)]^2}{2\sigma^2}} \tag{5.16}$$

The joint density of all the variables can then be written as

$$p(x, y, \theta, \sigma) = p(y, \theta, \sigma) \prod_t p(x(t)|y(t), \theta, \sigma) \tag{5.17}$$

Usually also the probability $p(y, \theta, \sigma)$ is stated in a factorisable form making the full joint probability density $p(x, y, \theta, \sigma)$ a product of many simple terms.

In supervised learning, the sequence y is assumed to be fully known, including any future data, which means that the full joint probability $p(x, y, \theta, \sigma)$ is not needed, only

$$p(x, \theta, \sigma|y) = p(x|y, \theta, \sigma)p(\theta, \sigma|y) \tag{5.18}$$

Typically y is assumed to be independent of θ and σ, i.e., $p(y, \theta, \sigma) = p(y)p(\theta, \sigma)$. This also means that $p(\theta, \sigma|y) = p(\theta, \sigma)$ and thus only $p(x|y, \theta, \sigma)$, given by the generative model equation 5.14, and the prior for the parameters $p(\theta, \sigma)$ is needed in supervised learning.

If the probability $p(y)$ is not modelled in supervised learning, it is impossible to treat missing elements of the sequence y. If the probability $p(y)$ is modelled, however, there are no problems. The posterior density is computed for all unknown variables, including the missing elements of y. In fact, unsupervised learning can be seen as a special case where the whole sequence y is unknown. In a probabilistic framework, the treatment of any missing values is possible as long as the model defines the joint density of all the variables in the model. It is, for instance, easy to treat missing elements of sequence x or mix freely between supervised and unsupervised learning depending on how large a part of the sequence y is known.

5.5.1   Priors and Hyperpriors

In the above model, the parameters θ and σ need to be assigned a prior probability and, in a more general situation, typically a prior probability for the model structure would also be needed.

In general, prior probabilities for variables should reflect the belief one has about the variables. As people may have difficulties in articulating these beliefs explicitly, some rules of thumb have been developed.

A lot of research has been conducted on uninformative priors. The term means that in some sense the prior gives as little information as possible about the value of the parameter, and as such is a good reference prior although with complex models it is usually impractical to compute the exact form of the uninformative prior.

Roughly speaking, uninformative priors can be defined by saying that in a parameterisation where moving a given distance in parameter space always corresponds to a similar change in the probability distribution the model defines, the parameters are uniformly distributed.

A simple example is provided by a Gaussian distribution parameterised by mean μ and variance σ^2. Doubling the variance always results in a qualitatively similar change in the distribution. Similarly, taking a step of size σ in the mean always corresponds to a similar change in the distribution. Reparametrisation by $\mu' = \mu/\sigma$ and $v = \ln \sigma$ will give a parametrisation where equal changes in parameters correspond to equal changes in distribution. A uniform prior on μ' and v would correspond to the prior $p(\mu, \sigma) \propto 1/\sigma^2$ in the original parameter space. If there is additional knowledge that μ should be independent of σ, then μ and v give the needed parameters and the prior is $p(\mu) \propto 1$ and $p(\sigma) \propto 1/\sigma$.

None of the above uninformative priors can actually be used because they are improper, meaning that they are not normalisable. This can easily be seen by considering a uniform distribution between $\pm\infty$. These priors are nevertheless good references and hint at useful parameterisations for models.

Often it is possible to utilise the fact that the model has a set of parameters which have a similar role. It is, for instance, reasonable to assume that all biases in an MLP network have a similar distribution. This knowledge can be utilised by modelling the distribution by a common parametrised distribution. Then the prior needs to be determined to these common parameters, called hyperparameters; as they control the distribution of a set of parameters, there should be fewer hyperparameters than parameters. The process can be iterated until the structural knowledge has been used. In the end there are usually only a few priors to determine and since there is typically a lot of data, these priors are usually not significant for the learining process.

For some, it may be helpful to think about the prior in terms of coding. By using the formula $L(x) = -\ln p(x)$, any probabilites can be translated into encoding. In coding terms, the prior means the aspects of the encoding which the sender and the receiver have agreed upon prior to the transmission of data.

5.6   Examples

5.6.1   Fixed Form Q

Let us model a set of observations, $x = x(1), \ldots, x(t)$, by a Gaussian distribution parameterised by mean m and log-std $v = \ln \sigma$. We shall approximate the posterior distribution by $q(m, v) = q(m)q(v)$, where both $q(m)$ and $q(v)$ are Gaussian. The parametrisation with log-std is chosen because the posterior of v is closer to Gaussian than the posterior of σ or σ^2 would be. (Notice that the parameterisation yielding close to Gaussian posterior distributions is connected to the uninformative priors discussed in section 5.5.1.)

Let the priors for m and v be Gaussian with means μ_m and μ_v and variances σ_m^2 and σ_v^2, respectively. The joint density of the observations and the parameters m and v is

$$p(x, m, v) = \left[\prod_t p(x(t)|m, v) \right] p(m)p(v) \tag{5.19}$$

As we can see, the posterior is a product of many simple terms.

Let us denote by $\bar{m}$ and $\tilde{m}$ the posterior mean and variance of m.

$$q(m; \bar{m}, \tilde{m}) = \frac{1}{\sqrt{2\pi\tilde{m}}} e^{-\frac{(m-\bar{m})^2}{2\tilde{m}}} \tag{5.20}$$

The distribution $q(v; \bar{v}, \tilde{v})$ is analogous.

The cost function is now

$$C_{m,v}(x) = \int q(m, v) \ln \frac{q(m, v)}{p(x, m, v)} dm dv =$$

$$\int q(m, v) \ln q(m) dm dv + \int q(m, v) \ln q(v) dm dv +$$

$$-\sum_{t=1}^{N} \int q(m, v) \ln p(x(t)|m, v) dm dv +$$

$$-\int q(m, v) \ln p(m) dm dv - \int q(m, v) \ln p(v) dm dv \tag{5.21}$$

We see that the cost function has many terms, all of which are expectations over $q(m, v)$. Since the approximation $q(m, v)$ is assumed to be factorised into $q(m, v) = q(m)q(v)$, it is fairly easy to compute these expectations. For instance, integrating the term $\int q(m, v) \ln q(m) dm dv$ over v yields $\int q(m) \ln q(m) dm$, since $\ln q(m)$ does not depend on v. Since $q(m)$ is assumed to be Gaussian with mean $\bar{m}$ and variance $\tilde{m}$, this integral is, in fact, minus the entropy of a Gaussian distribution and we have

$$\int q(m) \ln q(m) dm = -\frac{1}{2}(1 + \ln 2\pi\tilde{m}) \tag{5.22}$$

A similar term, with $\tilde{m}$ replaced by $\tilde{v}$, comes from $\int q(m,v) \ln q(v) dm dv$.

The terms where expectation is taken over $-\ln p(m)$ and $-\ln p(v)$ are also simple since

$$-\ln p(m) = \frac{1}{2} \ln 2\pi\sigma_m^2 + \frac{(m - \mu_m)^2}{2\sigma_2} \tag{5.23}$$

which means that we only need to be able to compute the expectation of $(m - \mu_m)^2$ over the Gaussian $q(m)$ having mean $\bar{m}$ and variance $\tilde{m}$. This yields

$$E\{(m - \mu_m)^2\} = E\{m^2\} - 2E\{m\mu_m\} + E\{\mu_m^2\} =$$
$$\bar{m}^2 + \tilde{m} - 2\bar{m}\mu_m + \mu_m^2 = (\bar{m} - \mu_m)^2 + \tilde{m} \tag{5.24}$$

since the variance can be defined by $\tilde{m} = E\{m^2\} - E\{m\}^2 = E\{m^2\} - \bar{m}^2$ which shows that $E\{m^2\} = \bar{m}^2 + \tilde{m}$. Integrating the equation 5.23 and substituting equation 5.24 thus yields

$$-\int q(m) \ln p(m) dm = \frac{1}{2} \ln 2\pi\sigma_m^2 + \frac{(\bar{m} - \mu_m)^2 + \tilde{m}}{2\sigma_2} \tag{5.25}$$

A similar term, with m replaced by v, will be obtained from $-\int q(v) \ln p(v) dv$.

The last terms are of the form $-\int q(m,v) \ln p(x(t)|m,v) dm dv$. Again we will find out that the factorisation $q(m,v) = q(m)q(v)$ simplifies the computation of these terms. Recall that $x(t)$ was assumed to be Gaussian with mean m and variance e^{2v}. The term over which the expectation is taken is thus

$$-\ln p(x(t)|m,v) = \frac{1}{2} \ln 2\pi + v + (x(t) - m)^2 e^{-2v} \tag{5.26}$$

The expectation over the term $(x(t) - m)^2 e^{-2v}$ is easy to compute since m and v are assumed to be posteriorly independent. This means that it is possible to take the expectation separately from $(x(t) - m)^2$ and e^{-2v}. Using a similar derivation as in equation 5.24 yields $(x(t) - \bar{m})^2 + \tilde{m}$ for the first term. The expectation over e^{-2v} is also fairly easy to compute:

$$\int q(v) e^{-2v} dv = \frac{1}{\sqrt{2\pi\tilde{v}}} \int e^{-\frac{(v-\bar{v})^2}{2\tilde{v}}} e^{-2v} dv = \frac{1}{\sqrt{2\pi\tilde{v}}} \int e^{-\frac{(v-\bar{v})^2 + 4v\tilde{v}}{2\tilde{v}}} dv =$$
$$\frac{1}{\sqrt{2\pi\tilde{v}}} \int e^{-\frac{v^2 - 2v\bar{v} + \bar{v}^2 + 4v\tilde{v}}{2\tilde{v}}} dv =$$
$$\frac{1}{\sqrt{2\pi\tilde{v}}} \int e^{-\frac{[v+(2\tilde{v}-\bar{v})]^2 + 4\bar{v}\tilde{v} - 4\tilde{v}^2}{2\tilde{v}}} =$$
$$\frac{1}{\sqrt{2\pi\tilde{v}}} \int e^{-\frac{[v+(2\tilde{v}-\bar{v})]^2}{2\tilde{v}}} e^{2\tilde{v} - 2\bar{v}} dv = e^{2\tilde{v} - 2\bar{v}} \tag{5.27}$$

This shows that taking expectation over equation 5.26 yields a term

$$-\int q(m,v)\ln p(x(t)|m,v)dmdv = \frac{1}{2}\ln 2\pi + \bar{v} + [(x(t)-\bar{m})^2 + \tilde{m}]e^{2\tilde{v}-2\bar{v}}$$
(5.28)

Collecting together all the terms, we obtain the following cost function

$$C_{m,v}(x;\bar{m},\tilde{m},\bar{v},\tilde{v}) = \sum_{t=1}^{N}\frac{1}{2}[(x(t)-\bar{m})^2 + \tilde{m}]e^{2\tilde{v}-2\bar{v}} + N\bar{v} +$$

$$\frac{(\bar{m}-\mu_m)^2 + \tilde{m}}{2\sigma_m^2} + \frac{(\bar{v}-\mu_v)^2 + \tilde{v}}{2\sigma_v^2} + \ln \sigma_m \sigma_v +$$

$$\frac{N}{2}\ln 2\pi - \frac{1}{2}\ln \tilde{m}\tilde{v} - 1$$
(5.29)

Assuming σ_m^2 and σ_v^2 are very large, the minimum of the cost function can be solved by setting the gradient of the cost function C to zero. This yields the following:

$$\bar{m} = \frac{1}{N}\sum_t x(t)$$
(5.30)

$$\tilde{m} = \frac{\sum_t(x(t)-\bar{m})^2}{N(N-1)}$$
(5.31)

$$\bar{v} = \frac{1}{2N} + \frac{1}{2}\ln\frac{1}{N-1}\sum_t(x(t)-\bar{m})^2$$
(5.32)

$$\tilde{v} = \frac{1}{2N}$$
(5.33)

If σ_m^2 and σ_v^2 cannot be assumed very large, the equations for $\bar{v}$ and $\tilde{v}$ are not that simple, but the solution can still be obtained by solving the zero-point of the gradient.

Figure 5.3 shows a comparison of the true and the approximate posterior distributions. The contours in both distributions are centred in the same region, a model that underestimates m. The contours for the two distributions are qualitatively similar although the true distribution is not symmetrical about the mean value of v.

5.6.2   Free Form Q

Instead of assuming the form for the approximate posterior distribution we could instead derive the optimal separable distribution (the functions that minimise the cost function subject to the constraint that they are normalised).

Instead of learning the log-std $v = \ln \sigma$, we shall learn the inverse noise variance $\beta = \sigma^{-2}$. The prior on β is assumed to be a Gamma distribution of

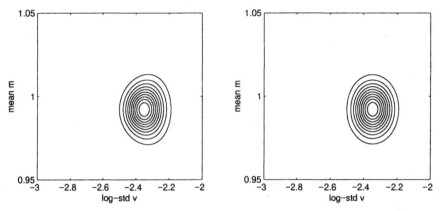

Fig. 5.3. Comparison of the true and approximate posterior distributions for a test set containing 100 data points drawn from a model with $m = 1$ and $\sigma = 0.1$. The plot on the left shows the true posterior distribution over m and v. The plot on the right shows the approximate posterior distribution consisting of a diagonal Gaussian distribution.

the form

$$p(\beta) = \frac{1}{\Gamma(c_\beta)} b_\beta^{c_\beta} \beta^{(c_\beta - 1)} \exp(-b_\beta \beta) \qquad (5.34)$$

Setting $b_\beta = c_\beta = 10^{-3}$ leads to a broad prior in $\ln \beta$. This is equivalent to assuming that σ_v is large in the log-std parameterisation.

The cost function that must be minimised is now

$$\begin{aligned} C &= D(q(x, m, \beta) || P(x|m, \beta)) - \ln P(m, \beta) \\ &= \int q(m, \beta) \ln \frac{q(m, \beta)}{P(x, m, \beta)} dm d\beta \end{aligned} \qquad (5.35)$$

If we assume a separable posterior, that is $q(m, \beta) = q(m)q(\beta)$ and substitute our priors into the cost function we obtain

$$\begin{aligned} C &= \int q(m)q(\beta) \ln \frac{q(m)q(\beta)}{[\prod_t P(x(t)|m, \beta)] P(m)P(\beta)} dm d\beta \\ &= \int q(m)q(\beta) \left[\ln q(m) + \ln q(\beta) - \ln P(m) - \ln P(\beta) \right. \\ &\qquad \left. - \sum_t \left(\frac{1}{2} \ln \frac{\beta}{2\pi} - \frac{\beta (x(t) - m)^2}{2} \right) \right] dm d\beta \end{aligned} \qquad (5.36)$$

Assuming we know $q(\beta)$ we can integrate over β in C (dropping any terms that are independent of m) to obtain

$$C = \int q(m) \left[\ln q(m) - \ln P(m) - \sum_t \left(-\frac{\bar{\beta} (x(t) - m)^2}{2} \right) \right] dm \qquad (5.37)$$

where $\bar{\beta}$ is the average value of β under the distribution $q(\beta)$. We can optimise the cost function with respect to $q(m)$ by performing a functional derivative.

$$\frac{\partial C}{\partial q(m)} = 1 + \ln q(m) - \ln P(m) - \sum_t \left(-\frac{\bar{\beta}\,(x(t) - m)^2}{2} \right) + \lambda_m$$
$$= 0 \tag{5.38}$$

where λ_m is a Lagrange multiplier introduced to ensure that $q(m)$ is normalised. Rearranging we see that

$$\ln q(m) = -1 - \frac{1}{2}\ln 2\pi\sigma_m^2 - \frac{(m - mu_m)^2}{2\sigma_m^2}$$
$$= +\sum_t \left(-\frac{\bar{\beta}\,(x(t) - m)^2}{2} \right) + \lambda_m \tag{5.39}$$

and so the approximate posterior distribution is a Gaussian with variance $\tilde{m} = \left(\sigma_m^{-2} + T\bar{\beta}\right)^{-1}$ and mean $\bar{m} = \tilde{m}\left(\frac{\mu_m}{\sigma_m^2} + \bar{\beta}\sum_t x(t)\right)$.

We can obtain the optimum form for $q(\beta)$ by marginalising the cost function over m and dropping terms independent of β.

$$C = \int q(\beta)\,[\ln q(\beta) - \ln P(\beta)$$
$$- \sum_t \left(\frac{1}{2}\ln\beta - \frac{\beta\left((x(t) - \bar{m})^2 + \tilde{m}\right)}{2} \right) \Bigg]\,d\beta \tag{5.40}$$

Again we can perform a functional derivative to obtain

$$\frac{\partial C}{\partial q(\beta)} = 1 + \ln q(\beta) - \ln P(\beta)$$
$$- \sum_t \left(\frac{1}{2}\ln\beta - \frac{\beta\left((x(t) - \bar{m})^2 + \tilde{m}\right)}{2} \right) + \lambda_\beta$$
$$= 0 \tag{5.41}$$

and so

$$\ln q(\beta) = -1 + \sum_t \left(\frac{1}{2}\ln\beta - \frac{\beta\left((x(t) - \bar{m})^2 + \tilde{m}\right)}{2} \right) + \ln P(\beta) + \lambda_\beta$$
$$\tag{5.42}$$

So the optimal posterior distribution is a Gamma distribution, with parameters $\hat{b}_\beta = b_\beta + \frac{\sum_t \left((x(t) - \bar{m})^2 + \tilde{m}\right)}{2}$ and $\hat{c}_\beta = c_\beta + \frac{T}{2}$. Therefore the expectation of β under the posterior distribution is $\bar{\beta} = \frac{\hat{c}_\beta}{\hat{b}_\beta}$.

The optimal distributions for m and β depend on each other ($q(m)$ is a function of $\bar{\beta}$ and $q(\beta)$ is a function of $\bar{m}$ and $\tilde{m}$) so the optimal solutions can be found by iteratively updating $q(m)$ and $q(\beta)$.

A general point is that the free form optimisation of the cost function will typically lead to a set of iterative update equations where each distribution is updated on the basis of the other distributions in the approximation.

We can also see that, if the parameterisation of the model is chosen appropriately, the optimal separable model has a similar form to the prior model. If the prior distributions are Gaussians, the posterior distributions are also Gaussians (likewise for Gamma distributions). If this is the case then we can say that we have chosen conjugate priors.

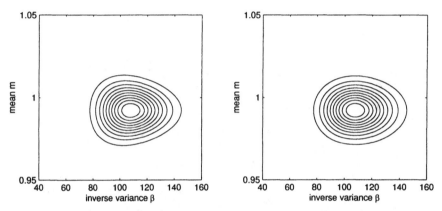

Fig. 5.4. Comparison of the true and approximate posterior distributions for a test set containing 100 data points drawn from a model with $m = 1$ and $\sigma = 0.1$. The plot on the left shows the true posterior distribution over m and β. The plot on the right shows the approximate posterior distribution derived by obtaining the optimal free form separable distribution.

Figure 5.4 shows a comparison of the true and the approximate posterior distributions. The contours in both distributions are centred in the same region, a model that underestimates m. The contours for the two distributions are qualitatively similar; the approximate distribution also shows the assymmetric density.

5.7 Conclusion

In ensemble learning, the search for good models is sensitive to high probability mass and so the problems of over - fitting inherent to maximum likelihood and maximum posterior probability methods are removed.

The approximation of the posterior distribution assumes some degree of factorisation of the true distribution in order to make the approximation more tractable. Additionally the fixed form approximation also assumes some functional form of the factors.

It is often possible to affect the correctness of the approximation by the choice of parameterisation of the model. Also, the learning process tries to make the approximation more correct.

The free form approximation of the separable distribution will often result in Gaussian, Gamma, or other distributions if the parameterisation of the model is chosen suitably. Therefore the optimisation process will suggest a parameterisation for the problem.

References

1. G. E. Hinton and D. van Camp. Keeping neural networks simple by minimizing the description length of the weights. In *Proceedings of the COLT'93*, pp. 5–13, Santa Cruz, California, 1993.
2. S. Kullback and R. A. Leible. On information and sufficiency. The Annals of Mathematical Statistics **22**: 79–86, 1951.
3. R M. Neal. Bayesian Learning for Neural Networks. Lecture Notes in Statistics No. 118. Springer-Verlag, 1996.
4. R M. Neal and G E. Hinton. A view of the EM algorithm that justifies incremental, sparse and other variants. In *Learning in Graphical Models*. M. I. Jordan, editor, 1998.
5. C E. Shannon. A mathematical theory of communication. Bell Systems Technical Journal **27**: 379–423 and 623–656, 1948.
6. S. M. Stigler. Translation of Laplace's 1774 memoir on "Probability of causes". Statistical Science, **1** (3): 359–378, 1986.

6 Bayesian Non-Linear Independent Component Analysis by Multi-Layer Perceptrons

Harri Lappalainen and Antti Honkela

6.1 Introduction

In this chapter, a non-linear extension to independent component analysis is developed. The non-linear mapping from source signals to observations is modelled by a multi-layer perceptron network and the distributions of source signals are modelled by mixture-of-Gaussians. The observations are assumed to be corrupted by Gaussian noise and therefore the method is more adequately described as non-linear independent factor analysis. The non-linear mapping, the source distributions and the noise level are estimated from the data. Bayesian approach to learning avoids problems with overlearning which would otherwise be severe in unsupervised learning with flexible non-linear models.

The linear principal and independent component analysis (PCA and ICA) model the data as having been generated by independent sources through a linear mapping. The difference between the two is that PCA restricts the distribution of the sources to be Gaussian, whereas ICA does not, in general, restrict the distribution of the sources.

In this chapter we introduce non-linear counterparts of PCA and ICA where the generative mapping from sources to data is not restricted to being linear. The general form of the models discussed here is

$$x(t) = f(s(t)) + n(t) \tag{6.1}$$

The vectors $x(t)$ are observations at time t, $s(t)$ are the sources and $n(t)$ the noise. The function $f()$ is a parameterised mapping from source space to observation space. It can be viewed as a model about how the observations were generated from the sources.

In the same way as their linear counterparts, the non-linear versions of PCA and ICA can be used in dimension reduction and feature extraction. The difference between linear and non-linear PCA is depicted in figure 6.1. In the linear PCA the data is described with a linear coordinate system whereas in the non-linear PCA the coordinate system is non-linear. The non-linear PCA and ICA can be used for similar tasks as their linear counterparts, but they can be expected to capture the structure of the data better if the data points lie in a non-linear manifold instead of a linear subspace.

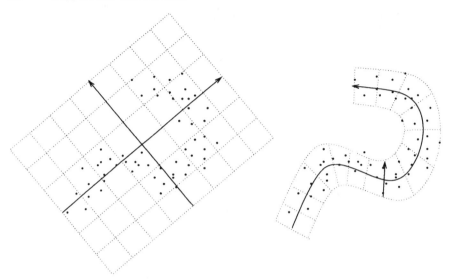

Fig. 6.1. On the left hand side the data is described with a linear coordinate system. On the right hand side the coordinate system is non-linear

Usually the linear PCA and ICA models do not have an explicit noise term $n(t)$ and the model is thus simply

$$x(t) = f(s(t)) = As(t) + b \tag{6.2}$$

The corresponding PCA and ICA models which include the noise term are often called factor analysis and independent factor analysis (FA and IFA) models. The non-linear models discussed here can therefore also be called non-linear factor analysis and non-linear independent factor analysis models.

In this chapter, the distribution of sources is modelled with Gaussian density in PCA and mixture-of-Gaussians density in ICA. Given enough Gaussians in the mixture, any density can be modelled with arbitrary accuracy using the mixture-of-Gaussians density, which means that the source density model is universal. Likewise, the non-linear mapping $f()$ is modelled by a multi-layer perceptron (MLP) network which can approximate any non-linear mapping with arbitrary accuracy given enough hidden neurons.

The noise on each observation channel (component of data vectors) is assumed to be independent and Gaussian, but the variance of the noise on different channels is not assumed to be equal. The noise could be modelled with a more general distribution, but we shall restrict the discussion to the simple Gaussian case. After all, noise is supposed to be something uninteresting and unstructured. If the noise is not Gaussian or independent, it is a sign of interesting structure which should be modelled by the generative mapping from the sources.

6.2   Choosing Among Competing Explanations

Each model with particular values for sources, parameters and noise terms can be considered as an explanation for the observations. Even with linear PCA and ICA there are infinitely many possible models which explain the observations completely. With flexible non-linear models like an MLP network, the number of possible explanations is — loosely speaking — even higher (although mathematically speaking, ∞^2 would still be 'only' ∞).

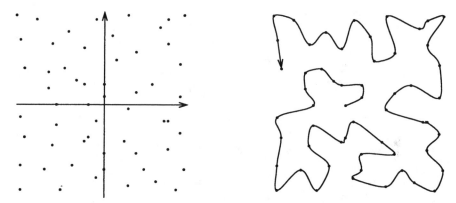

Fig. 6.2. The data is generated by two independent evenly distributed sources as shown on the left. Given enough hidden neurons, an MLP network is able to model the data as having been generated by a single source through a very non-linear mapping, depicted on the right.

An example of competing explanations is given in figure 6.2. The data is sampled from an even distribution inside a square. This is equivalent to saying that two independent sources, each evenly distributed, have generated the data as shown on the left-hand side of the figure. If we only look at the probability of the data, the non-linear mapping depicted on the right hand side of the figure is an even better explanation as it gives very high probabilities to exactly those data points that actually occurred. However, it seems intuitively clear that the non-linear model in figure 6.2 is much more complex than the available data would justify.

The exact Bayesian solution is that instead of choosing a single model, all models are used by weighting them according to their posterior probabilities. In other words, each model is taken into account in proportion with how probable it seems in light of the observations.

If we look at the predictions the above two models give about future data points, we notice that the more simple linear model with two sources predicts new points inside the square but the more complex non-linear model with one source predicts new points only along the curved line. The prediction given by the more simple model is evidently closer to the prediction obtained by the exact Bayesian approach where the predictions of all models would be taken into account by weighting them according to the posterior probabilities of the models.

With complex non-linear models like MLP networks, the exact Bayesian treatment is computationally intractable and we resort to ensemble learning, which is discussed in chapter 5. In ensemble learning, a computationally tractable parametric approximation is fitted to the posterior probabilities of the models.

Section 5.4.2 shows that ensemble learning can be interpreted as finding the most simple explanation for the observations.[1] This agrees with the intuition that in figure 6.2, the simple linear model is better than the complex non-linear model.

The fact that we are interested in simple explanations also explains why non-linear ICA is needed at all if we can use non-linear PCA. The non-linearity of the mapping allows the PCA model to represent any time independent probability density of the observations as originating from independent sources with Gaussian distributions. It would therefore seem that the non-Gaussian source models used in the non-linear ICA cannot further increase the representational power of the model. However, for many naturally occurring processes the representation with Gaussian sources requires more complex non-linear mappings than the representation with mixtures-of-Gaussians. Therefore the non-linear ICA will often find a better explanation for the observations than the non-linear PCA.

Similar considerations also explain why we use the MLP network for modelling the non-linearity. Experience has shown that with MLP networks it is easy to model fairly accurately many naturally occurring multi-dimensional processes. In many cases the MLP networks give a more simple parameterisation for the non-linearity than, for example, Taylor or Fourier series expansions.

On the other hand, it is equally clear that the ordinary MLP networks with sigmoidal non-linearities are not the best models for all kinds of data. With the ordinary MLP networks it is, for instance, difficult to model mappings which have products of the sources. The purpose of this chapter is not to give the ultimate model for any data but rather to give a good model for many data, from which one can start building more sophisticated models by incorporating domain-specific knowledge. Most notably, the source models described here do not assume time-dependencies, which are often significant.

---

[1] The complexity of the explanation is defined as the number of bits it takes to encode the observation using the model. In this case, one would measure the total code length of the sources $s(t)$, the parameters of the mapping and the noise $n(t)$.

6.3   Non-Linear Factor Analysis

This section introduces a non-linear counterpart of principal component analysis. As explained in section 6.1, the model includes a noise term and we shall therefore call it non-linear factor analysis. Learning is based on Bayesian ensemble learning which was introduced in chapter 5. In order to keep the derivations simple, only Gaussian probability distributions are used which allows us to utilise many of the formulas derived in section 5.6.1.

The posterior probability density of the unknown variables is approximated by a Gaussian distribution. As in chapter 5, the variances of the Gaussian distributions of the model are parameterised by logarithm of standard deviation, log-std, because then the posterior distribution of these parameters will be closer to Gaussian which then agrees better with the assumption that the posterior is Gaussian.

6.3.1   Definition of the Model

Sources *s(t)*

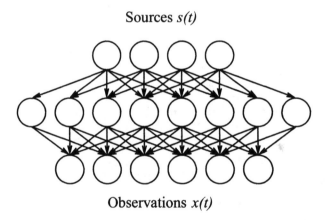

Observations *x(t)*

Fig. 6.3. The mapping from sources to observations is modelled by the familiar MLP network. The sources are on the top layer and observations in the bottom layer. The middle layer consists of hidden neurons each of which computes a non-linear function of the inputs.

The schematic structure of the mapping is shown in figure 6.3. The non-linearity of each hidden neuron is the hyperbolic tangent, which is the same as the usual logistic sigmoid except for a scaling. The equation defining the mapping is

$$\boldsymbol{x}(t) = \boldsymbol{f}(\boldsymbol{s}(t)) + \boldsymbol{n}(t) = \mathrm{B}\tanh(\mathrm{A}\boldsymbol{s}(t) + \boldsymbol{a}) + \boldsymbol{b} + \boldsymbol{n}(t) \tag{6.3}$$

The matrices A and B are the weights of first and second layer and a and b are the corresponding biases.

The noise is assumed to be independent and Gaussian and therefore the probability distribution of $x(t)$ is

$$x(t) \sim N(f(s(t)), e^{2v_x}) \tag{6.4}$$

Each component of the vector v_x gives the log-std of the corresponding component of $x(t)$.

The sources are assumed to have zero mean Gaussian distributions and again the variances are parameterised by log-std v_s.

$$s(t) \sim N(0, e^{2v_s}) \tag{6.5}$$

Since the variance of the sources can vary, variance of the weights A on the first layer can be fixed to a constant, which we choose to be one, without losing any generality from the model. This is not case for the second layer weights. Due to the non-linearity, the variances of the outputs of the hidden neurons are bounded from above and therefore the variance of the second layer weights cannot be fixed. In order to enable the network to shut off extra hidden neurons, the weights leaving one hidden neuron share the same variance parameter.[2]

$$B \sim N(0, e^{2v_B}) \tag{6.6}$$

The elements of the matrix B are assumed to have a zero mean Gaussian distribution with individual variances for each column and thus the dimension of the vector v_B is the number of hidden neurons. Both biases a and b have Gaussian distributions parameterised by mean and log-std.

The distributions are summarised in equations 6.7 – 6.12.

$$x(t) \sim N(f(s(t)), e^{2v_x}) \tag{6.7}$$
$$s(t) \sim N(0, e^{2v_s}) \tag{6.8}$$
$$A \sim N(0, 1) \tag{6.9}$$
$$B \sim N(0, e^{2v_B}) \tag{6.10}$$
$$a \sim N(m_a, e^{2v_a}) \tag{6.11}$$
$$b \sim N(m_b, e^{2v_b}) \tag{6.12}$$

The distributions of each set of log-std parameters are modelled by Gaussian distributions whose parameters are usually called hyperparameters.

$$v_x \sim N(m_{v_x}, e^{2v_{v_x}}) \tag{6.13}$$
$$v_s \sim N(m_{v_s}, e^{2v_{v_s}}) \tag{6.14}$$
$$v_B \sim N(m_{v_B}, e^{2v_{v_B}}) \tag{6.15}$$

---

[2] A hidden neuron will be shut off if all leaving weights are close to zero. Thinking in coding terms, it is easier for the network to encode this in one variance parameter than to encode it independently for all the weights.

The priors of m_a, v_a, m_b, v_b and the six hyperparameters $m_{v_s}, \ldots, v_{v_B}$ are assumed to be Gaussian with zero mean and standard deviation 100, i.e., the priors are assumed to be very flat.

6.3.2   Cost Function

In ensemble learning, the goal is to approximate the posterior pdf of all the unknown values in the model. Let us denote the observations by X. Everything else in the model is unknown, i.e., the sources, parameters and hyperparameters. Let us denote all these unknowns by the vector $\boldsymbol{\theta}$. The cost function measures the misfit between the actual posterior pdf $p(\boldsymbol{\theta}|X)$ and its approximation $q(\boldsymbol{\theta}|X)$.

The posterior is approximated as a product of independent Gaussian distributions

$$q(\boldsymbol{\theta}|X) = \prod_i q(\theta_i|X) \tag{6.16}$$

Each individual Gaussian $q(\theta_i|X)$ is parameterised by the posterior mean $\bar{\theta}_i$ and variance $\tilde{\theta}_i$ of the parameter.

The functional form of the cost function $C_{\boldsymbol{\theta}}(X; \bar{\boldsymbol{\theta}}, \tilde{\boldsymbol{\theta}})$ is given in chapter 5. The cost function can be interpreted to measure the misfit between the actual posterior $p(\boldsymbol{\theta}|X)$ and its factorial approximation $q(\boldsymbol{\theta}|X)$. It can also be interpreted as measuring the number of bits it would take to encode X when approximating the posterior pdf of the unknown variables by $q(\boldsymbol{\theta}|X)$.

The cost function is minimised with respect to the posterior means $\bar{\theta}_i$ and variances $\tilde{\theta}_i$ of the unknown variables θ_i. The end result of the learning is therefore not just an estimate of the unknown variables, but a distribution over the variables.

The simple factorising form of the approximation $q(\boldsymbol{\theta}|X)$ makes the cost function computationally tractable. The cost function can be split into two terms, C_q and C_p, where the former is an expectation over $\ln q(\boldsymbol{\theta}|X)$ and the latter is an expectation over $-\ln p(X, \boldsymbol{\theta})$.

It turns out that the term C_q is not a function of the posterior means $\bar{\theta}_i$ of the parameters, only the posterior variances. It has a similar term for each unknown variable.

$$C_q(X; \tilde{\boldsymbol{\theta}}) = \sum_i -\frac{1}{2} \ln 2\pi e \tilde{\theta}_i \tag{6.17}$$

Most of the terms of C_p are also trivial. The Gaussian densities in equations 6.8 – 6.15 yield terms of the form

$$-\ln p(\theta) = \frac{1}{2}(\theta - m_\theta)^2 e^{-2v_\theta} + v_\theta + \frac{1}{2}\ln 2\pi \tag{6.18}$$

Since θ, m_θ and v_θ are independent in q, the expectation over q yields

$$\frac{1}{2}[(\bar{\theta}-\bar{m}_\theta)^2+\tilde{\theta}+\tilde{m}_\theta]e^{2\tilde{v}_\theta-2\bar{v}_\theta}+\bar{v}_\theta+\frac{1}{2}\ln 2\pi\,. \qquad (6.19)$$

Only the term originating from equation 6.7 needs some elaboration. Equation 6.7 yields

$$-\ln p(x)=\frac{1}{2}(x-f)^2e^{-2v_x}+v_x+\frac{1}{2}\ln 2\pi \qquad (6.20)$$

and the expectation over q is

$$\frac{1}{2}[(x-\bar{f})^2+\tilde{f}]e^{2\tilde{v}_x-2\bar{v}_x}+\bar{v}_x+\frac{1}{2}\ln 2\pi \qquad (6.21)$$

The rest of this section is dedicated to evaluating the posterior mean $\bar{f}$ and variance $\tilde{f}$ of the function f. We shall begin from the sources and weights and show how the posterior mean and variance can be propagated through the network yielding the needed posterior mean and variance of the function f at the output. The effect of non-linearities g of the hidden neurons are approximated by first and second order Taylor's series expansions around the posterior mean. Apart from that, the computation is analytical.

The function f consists of two multiplications with matrices and a non-linearity between them. The posterior mean and variance for a product $u = yz$ are

$$\bar{u} = E\{u\} = E\{yz\} = E\{y\}E\{z\} = \bar{y}\bar{z} \qquad (6.22)$$

and

$$\tilde{u} = E\{u^2\} - \bar{u}^2 = E\{y^2z^2\} - (\bar{y}\bar{z})^2 =$$
$$E\{y^2\}E\{z^2\} - \bar{y}^2\bar{z}^2 = (\bar{y}^2 + \tilde{y})(\bar{z}^2 + \tilde{z}) - \bar{y}^2\bar{z}^2 \qquad (6.23)$$

given that y and z are posteriorly independent. According to the assumption of the factorising form of $q(\theta|X)$, the sources and the weights are independent and we can use the above formulas. The inputs going to hidden neurons consist of sums of products of weights and sources, each posteriorly independent, and it is therefore easy to compute the posterior mean and variance of the inputs going to the hidden neurons; both the means and variances of a sum of independent variables add up.

Let us now pick one hidden neuron having non-linearity g and input ξ, i.e., the hidden neuron is computing $g(\xi)$. At this point we are not assuming any particular form of g although we are going to use $g(\xi) = \tanh \xi$ in all the experiments; the following derivation is general and can be applied to any sufficiently smooth function g.

In order to be able to compute the posterior mean and variance of the function g, we are going to apply the Taylor's series expansion around the posterior mean $\bar{\xi}$ of the input. We choose the second order expansion when computing the mean and the first order expansion when computing the variance. The choice is purely practical; higher order expansions could be used as well but these are the ones that can be computed from the posterior mean and variance of the inputs alone.

$$\bar{g}(\xi) \approx g(\bar{\xi}) + \frac{1}{2}g''(\bar{\xi})\tilde{\xi} \tag{6.24}$$

$$\tilde{g}(\xi) \approx [g'(\bar{\xi})]^2 \tilde{\xi} \tag{6.25}$$

After having evaluated the outputs of the non-linear hidden neurons, it would seem that most of the work has already been done. After all, it was already shown how to compute the posterior mean and variance of a weighted sum and the outputs of the network will be weighted sums of the outputs of the hidden neurons. Unfortunately, this time the terms in the sum are no longer independent. The sources are posteriorly independent by virtue of the approximation $q(\theta|X)$, but the values of the hidden neurons are posteriorly dependent which enforces us to use a more complicated scheme for computing the posterior variances of these weighted sums. The posterior means will be as simple as before, though.

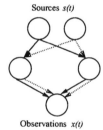

Sources s(t)

Observations x(t)

Fig. 6.4. The converging paths from two sources are shown. Both input neurons affect the output neuron through two paths going through the hidden neurons. This means that the posterior variances of the two hidden neurons are neither completely correlated nor uncorrelated and it is impossible to compute the posterior variance of the output neuron without keeping the two paths separate. Effectively this means computing the Jacobian matrix of the output with respect to the inputs.

The reason for the outputs of the hidden neurons to be posteriorly dependent is that the value of one source can potentially affect all the outputs. This is illustrated in figure 6.4. Each source affects the output of the whole network through several paths and in order to be able to determine the variance of the outputs, the paths originating from different sources need to be kept separate. This is done by keeping track of the partial derivatives $\frac{\partial g(\xi)}{\partial s_i}$. Equation 6.26 shows how the total posterior variance of the output $g(\xi)$ of one of the hidden neurons can be split into terms originating from each source plus a term $\tilde{g}^*(\xi)$ which contains the variance originating from the weights and biases, i.e., those variables which affect any one output through only a single path.

$$\tilde{g}(\xi) = \tilde{g}^*(\xi) + \sum_i \tilde{s}_i \left[\frac{\partial g(\xi)}{\partial s_i}\right]^2 \tag{6.26}$$

When the outputs are multiplied by weights, it is possible to keep track of how this affects the posterior mean, the derivatives w.r.t. the sources and the variance originating from other variables than the sources, i.e., from weights and biases. The total variance of the output of the network is then obtained by

$$\tilde{f} = \tilde{f}^* + \sum_i \tilde{s}_i \left[\frac{\partial f}{\partial s_i} \right]^2 \tag{6.27}$$

where f denotes the components of the output and we have computed the posterior variance of the outputs of the network which is needed in equation 6.21. To recapitulate what is done, the contributions of different sources to the variances of the outputs of the network are monitored by computing the Jacobian matrix of the outputs w.r.t. the sources and keeping this part separate from the variance originating from other variables.

The only approximations done in the computation are the ones approximating the effect of non-linearity. If the hidden neurons were linear, the computation would be exact. The non-linearity of the hidden neurons is dealt with by linearising around the posterior mean of the inputs of the hidden neurons. The smaller the variances the more accurate this approximation is. With increasing non-linearity and variance of the inputs, the approximation gets worse.

Compared to ordinary forward phase of an MLP network, the computational complexity is greater by about a factor of $5N$, where N is the number of sources. The factor five is due to propagating distributions instead of plain values. The need to keep the paths originating from different sources separate explains the factor N. Fortunately, much of the extra computation can be put to good use later on when adapting the distributions of variables.

6.3.3   Update Rules

Any standard optimisation algorithm could be used for minimising the cost function $C(\mathrm{X}; \bar{\boldsymbol{\theta}}, \tilde{\boldsymbol{\theta}})$ with respect to the posterior means $\bar{\boldsymbol{\theta}}$ and variances $\tilde{\boldsymbol{\theta}}$ of the unknown variables. As usual, however, it makes sense to use the particular structure of the function to be minimised.

Those parameters which are means or log-std of Gaussian distributions, e.g., m_b, m_{v_B}, v_a and v_{v_x}, can be solved in the same way as the parameters of Gaussian distribution were solved in section 5.1. Since the parameters have Gaussian priors, the equations do not have analytical solutions, but Newton iteration can be used. For each Gaussian, the posterior mean and variance of the parameter governing the mean is solved first by assuming all other variables constant and then the same thing is done for the log-std parameter, again assuming all other variables constant.

Since the mean and variance of the output of the network and thus also the cost function was computed layer by layer, it is possible to use the ordinary back-propagation algorithm to evaluate the partial derivatives of the part C_p of the cost function w.r.t. the posterior means and variances of the sources, weights and biases. Assuming the derivatives computed, let us first take a look at the posterior variances $\tilde{\theta}$.

The effect of the posterior variances $\tilde{\theta}$ of sources, weights and biases on the part C_p of the cost function is mostly due to the effect on $\tilde{f}$ which is usually very close to linear (this was also the approximation made in the evaluation of the cost function). The terms $\tilde{f}$ have a linear effect on the cost function, as is seen in equation 6.21, which means that the overall effect of the terms $\tilde{\theta}$ on C_p is close to linear. The partial derivative of C_p with respect to $\tilde{\theta}$ is therefore roughly constant and it is reasonable to use the following fixed point equation to update the variances:

$$0 = \frac{\partial C}{\partial \tilde{\theta}} = \frac{\partial C_p}{\partial \tilde{\theta}} + \frac{\partial C_q}{\partial \tilde{\theta}} = \frac{\partial C_p}{\partial \tilde{\theta}} - \frac{1}{2\tilde{\theta}} \Rightarrow \tilde{\theta} = \frac{1}{2\frac{\partial C_p}{\partial \tilde{\theta}}} . \tag{6.28}$$

The remaining parameters to be updated are the posterior means $\bar{\theta}$ of the sources, weights and biases. For those parameters it is possible to use Newton iteration since the corresponding posterior variances $\tilde{\theta}$ actually contain the information about the second order derivatives of the cost function C w.r.t. $\bar{\theta}$. It holds

$$\tilde{\theta} \approx \frac{1}{\frac{\partial^2 C}{\partial \bar{\theta}^2}} \tag{6.29}$$

and thus the step in Newton iteration can be approximated

$$\bar{\theta} \leftarrow \bar{\theta} - \frac{\frac{\partial C_p}{\partial \bar{\theta}}}{\frac{\partial^2 C}{\partial \bar{\theta}^2}} \approx \bar{\theta} - \frac{\partial C_p}{\partial \bar{\theta}} \tilde{\theta} \tag{6.30}$$

Equation 6.29 would be exact if the posterior pdf $p(\theta|X)$ were exactly Gaussian. This would be true if the mapping f were linear. The approximation in equation 6.29 is therefore good as long as the function f is roughly linear around the current estimate of $\bar{\theta}$.

Avoiding Problems Originating from Approximation of the Non-Linearity of the Hidden Neurons. The approximations in equations 6.24 and 6.25 can give rise to problems with ill defined posterior variances of sources or first layer weights A or biases a. This is because the approximations take into account only local behaviour of the non-linearities g of the hidden neurons. With MLP networks the posterior is typically multimodal and therefore, in a valley between two maxima, it is possible that the second order derivative of the logarithm of the posterior w.r.t. a parameter θ is positive. This means that the derivative of the C_p part of the cost function with respect to the posterior variance $\tilde{\theta}$ of that parameter is negative, leading to a negative estimate of variance in (6.28).

It is easy to see that the problem is due to the local estimate of g since the logarithm of the posterior eventually has to go to negative infinity. The derivative of the C_p term w.r.t. the posterior variance $\tilde{\theta}$ will thus be positive for large $\tilde{\theta}$, but the local estimate of g fails to account for this.

In order to discourage the network from adapting itself to areas of parameter space where the problems might occur and to deal with the problem if it does occur, the terms in equation 6.24 which give rise to negative derivative of C_p with respect to $\tilde{\theta}$ will be neglected in the computation of the gradients. As this can only make the estimate of $\tilde{\theta}$ in equation 6.28 smaller, this leads, in general, to increasing the accuracy of the approximations in equation 6.24 and equation 6.25.

Stabilising the Fixed-Point Update Rules. The adaptations rules in equations 6.28 and 6.30 assume other parameters to be constant. The weights, sources and biases are updated all at once, however, because it would not be computationally efficient to update only one at a time. The assumption of independence is not necessarily valid, particularly for the posterior means of the variables, which may give rise to instabilities. Several variables can have a similar effect on outputs and when they are all updated to the values which would be optimal given that the others stay constant, the combined effect is too large.

This type of instability can be detected by monitoring the directions of updates of individual parameters. When the problem of correlated effects occurs, consecutive updated values start oscillating. A standard way to dampen these oscillations in fixed point algorithms is to introduce a learning parameter α for each parameter and update it according to the following rule:

$$\alpha \leftarrow \begin{cases} 0.8\alpha & \text{if sign of change was different} \\ \min(1, 1.05\alpha) & \text{if sign of change was the same} \end{cases} \tag{6.31}$$

This gives the following fixed point update rules for the posterior means and variances of the sources, weights and the biases:

$$\bar{\theta} \leftarrow \bar{\theta} - \alpha_{\bar{\theta}} \frac{\partial C_p}{\partial \bar{\theta}} \tilde{\theta} \tag{6.32}$$

$$\tilde{\theta} \leftarrow \frac{1}{\left[2\frac{\partial C_p}{\partial \tilde{\theta}}\right]^{\alpha_{\tilde{\theta}}}} \tilde{\theta}^{1-\alpha_{\tilde{\theta}}} \tag{6.33}$$

The reason why a weighted geometric rather than arithmetic mean is applied to the posterior variances is that variance is a scale parameter. The relative effect of adding a constant to the variance varies with the magnitude of the variance whereas the relative effect of multiplying by a constant is invariant.

Using Additional Information to Update Sources. With sources, it is possible to measure and compensate for some of the correlated effects in the updates. Recall that the Jacobian matrix of the output f of the network w.r.t. the sources was computed when taking into account the effects of multiple paths of propagating the values of sources. This will be used to compensate the assumption of independent updates, in addition to the learning rates α.

Suppose we have two sources whose effects on outputs are positively correlated. Assuming the effects are independent means that the step will be too large and the actual step size should be less than what the Newton iteration suggests. This can be detected from computing the change resulting in the outputs and projecting it back for each source independently to see how much each source alone should change to produce the same change in the outputs. The difference between the change of one source in the update and the change resulting from all the updates can then be used to adjust the step sizes in the update.

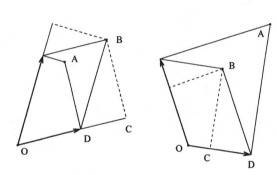

Fig. 6.5. Illustration of the correction of error resulting from assuming independent updates of the sources. The figures show the effect two sources have on the outputs. On the left-hand side the effects of sources on x are positively correlated and consequently the step sizes are overestimated. On the right-hand side the effects are negatively correlated and the step sizes are underestimated.

Two examples of correction are depicted in figure 6.5. The left-hand graph shows a case where the effects of sources on the outputs are positively correlated and the right-hand graph has negatively correlated effects. Current output of the network is in the origin O and the minimum of the cost function is in point A. Arrows show where the output would move if the sources were minimised independently. The combined updates would then take the output to point B.

As the effects of sources on x are correlated, point B, the resulting overall change in x, differs from point A. Projecting the point B back to the sources, comparison between the resulting step size C and the desired step size D can be used to adjust the step size. The new step size on the source would be D/C times the original. With positively correlated effects the adjusting factor D/C is less than one, but with negatively correlated sources it is greater than one. For the sake of stability, the corrected step is restricted to be at most twice the original.

6.4   Non-Linear Independent Factor Analysis

The non-linear factor analysis model introduced in the previous section has Gaussian distributions for the sources. In this section we are going to show how that model can easily be extended to have mixture-of-Gaussians source models. In doing so, we are largely following the method introduced in [1] for Bayesian linear independent factor analysis. The resulting model is a non-linear counterpart of ICA or, more accurately, a non-linear counterpart of independent factor analysis because the model includes finite noise. The difference between the models is similar to that between linear PCA and ICA because the first layer weight matrix A in the network has the same indeterminacies in non-linear PCA as in linear PCA. The indeterminacy is discussed in section 2.2.

According to the model for the distribution of the sources, there are several Gaussian distributions and at each time instant the source originates from one of them. Let us denote the index of the Gaussian from which the source $s_i(t)$ originates by $M_i(t)$. The model for the distribution for the ith source at time t is

$$p(s_i(t)|\theta) = \sum_{M_i(t)} P(M_i(t)|\theta)p(s_i(t)|\theta, M_i(t)) \tag{6.34}$$

where $p(s_i(t)|\theta, M_i(t) = j)$ is a time-independent Gaussian distribution with its own mean m_{ij} and log-std v_{ij}. The probabilities $P(M_i(t)|\theta)$ of different Gaussians are modelled with time-independent soft-max distributions.

$$P(M_i(t) = j|\theta) = \frac{e^{c_{ij}}}{\sum_{j'} e^{c_{ij'}}} \tag{6.35}$$

Each combination of Gaussians producing the sources can be considered a different model. The number of these models is enormous, of course, but their posterior distribution can still be approximated by a similar factorial approximation which is used for other variables.

$$Q(\mathrm{M}|\mathrm{X}) = \prod_{M_i(t)} Q(M_i(t)|\mathrm{X}) \tag{6.36}$$

Without losing any further generality, we can now write

$$q(s_i(t), M_i(t)|\theta) = Q(M_i(t)|\theta)q(s_i(t)|\theta, M_i(t)) \tag{6.37}$$

which yields

$$q(s_i(t)|\theta) = \sum_j q(M_i(t) = j|\theta)Q(s_i(t)|\theta, M_i(t) = j) \tag{6.38}$$

This means that the approximating ensemble for the sources has a form similar to the prior, i.e., an independent mixture of Gaussians, although the posterior mixture is different at different times.

Due to the assumption of factorial posterior distribution of the models, the cost function can be computed as easily as before. Let us denote $Q(M_i(t) = j|\theta) = \dot{s}_{ij}(t)$ and the posterior mean and variance of $q(s_i(t)|\theta, M_i(t) = j)$ by $\bar{s}_{ij}(t)$ and $\tilde{s}_{ij}(t)$. It easy to see that the posterior mean and variance of $s_i(t)$ are

$$\bar{s}_i(t) = \sum_j \dot{s}_{ij}(t)\bar{s}_{ij}(t) \tag{6.39}$$

$$\tilde{s}_i(t) = \sum_j \dot{s}_{ij}(t)[\tilde{s}_{ij}(t) + (\bar{s}_{ij}(t) - \bar{s}_i(t))^2] \tag{6.40}$$

After having computed the posterior mean $\bar{s}_i$ and variance $\tilde{s}_i$ of the sources, the computation of the C_p part of the cost function proceeds as with non-linear factor analysis. The C_q part yields terms of the form

$$q(s_i(t)|X)\ln q(s_i(t)|X) =$$

$$\sum_j \dot{s}_{ij}(t)q(s_i(t)|M_i(t) = j, X)\ln \sum_j \dot{s}_{ij}(t)q(s_i(t)|M_i(t), X) =$$

$$\sum_j \dot{s}_{ij}(t)q(s_i(t)|M_i(t) = j, X)\ln \dot{s}_{ij}(t)q(s_i(t)|M_i(t), X) =$$

$$\sum_j \dot{s}_{ij}(t)[\ln \dot{s}_{ij}(t) + q(s_i(t)|M_i(t) = j, X)\ln q(s_i(t)|M_i(t) = j, X)]$$

$$\tag{6.41}$$

and we have thus reduced the problem to a previously solved one. The terms $q(s_i(t)|M_i(t), X)\ln q(s_i(t)|M_i(t), X)$ are the same as for the non-linear factor analysis and otherwise the equation has the same form as in model selection in chapter 5. This is not surprising since the terms $Q(M_i(t)|X)$ are the probabilities of different models and we are, in effect, therefore doing factorial model selection.

Most update rules are the same as for non-linear factor analysis. Equations 6.39 and 6.40 bring the terms $\dot{s}_{ij}(t)$ for updating the means m_{ij} and log-std parameters v_{ij} of the sources. It turns out that they will both be weighted with $\dot{s}_{ij}$, i.e., the observation is used for adapting the parameters in proportion to the posterior probability of that observation originating from that particular Gaussian distribution.

6.5 Experiments

6.5.1 Learning Scheme

The learning scheme for all the experiments was the same. First, linear PCA is used to find sensible initial values for the posterior means of the sources. The method was chosen because it has given good results in initialising the model vectors of a self-organising map (SOM). The posterior variances of the sources are initialised to small values. Good initial values are important for the method since the network can effectively prune away unused parts as will be seen later. Initially the weights of the network have random values and the network has quite bad representation for the data. If the sources were adapted from random values also, the network would consider many of the sources useless for the representation and prune them away. This would lead to a local minimum from which the network would not recover.

Therefore the sources are fixed at the values given by linear PCA for the first 50 iterations through the whole data. This is long enough for the network to find a meaningful mapping from sources to the observations, thereby justifying using the sources for the representation. For the same reason, the parameters controlling the distributions of the sources, weights, noise and the hyperparameters are not adapted during the first 100 iterations. They are adapted only after the network has found sensible values for the variables whose distributions these parameters control.

In all simulations, the total number of iterations is 7500, where one iteration means going through all the observations. For non-linear independent factor analysis simulations, a non-linear subspace is estimated with 2000 iterations by the non-linear factor analysis after which the sources are rotated with a linear ICA algorithm. In these experiments, FastICA was used [4]. The rotation of the sources is compensated for by an inverse rotation to the first layer weight matrix A. The non-linear independent factor analysis algorithm is then applied for the remaining 5500 iterations.

6.5.2 Helix

Let us first look at a toy problem which shows that it is possible to find a non-linear subspace and model it with an MLP network in an unsupervised manner. A set of 1000 data points, shown on the left plot of figure 6.6, were generated from a normally distributed source s into a helical subspace. The z-axis had a linear correspondence to the source and the x- and y-axes were sine and cosine: $x = sin(\pi s)$, $y = cos(\pi s)$ and $z = s$. Gaussian noise with standard deviation 0.05 was added to all three data components.

One-dimensional non-linear subspaces were estimated with the non-linear independent factor analysis algorithm. Several different quantities of hidden neurons and initialisations of the MLP networks were tested and the network which minimised the cost function was chosen. The best network had 16 hidden neurons. The original noisy data and the means of the outputs of the best MLP network are shown in figure 6.6. It is evident that the network was able to learn the correct subspace. Only the tails of the helix are somewhat distorted. The network estimated the standard deviations of the noise on different data components to be 0.052, 0.055 and 0.050. This is in close correspondence with the actual noise level of 0.05.

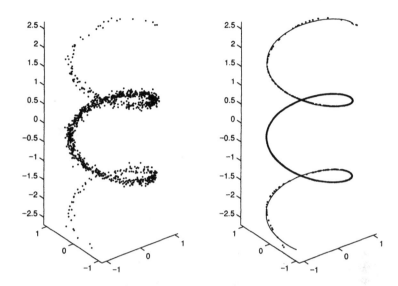

Fig. 6.6. The plot on the left shows the data points and the plot on the right shows the reconstructions made by the network together with the underlying helical subspace. The MLP network has clearly been able to find the underlying one-dimensional non-linear subspace where the data points lie.

This problem is not enough to demonstrate the advantages of the method since it does not prove that the method is able to deal with high dimensional latent spaces. This problem was chosen simply because it is easy to visualise.

6.5.3   Non-Linear Artificial Data

Gaussian Sources. The following experiments with non-linear factor analysis algorithms demonstrate the ability of the network to prune away unused parts. The data was generated from five normally distributed sources through a non-linear mapping. The mapping was generated by a randomly initialised MLP network having 20 hidden neurons and ten output neurons. Gaussian noise with standard deviation of 0.1 was added to the data. The non-linearity for the hidden neurons was chosen to be the inverse hyperbolic sine, which means that the non-linear factor analysis algorithm using MLP network with *tanh* non-linearities cannot use exactly the same weights.

Figure 6.7 shows how much of the energy remains in the data when a number of linear PCA components are extracted. This measure is often used to deduce the linear dimension of the data. As the figure shows, there is no obvious turn in the curve and it would be impossible to deduce the non-linear dimension.

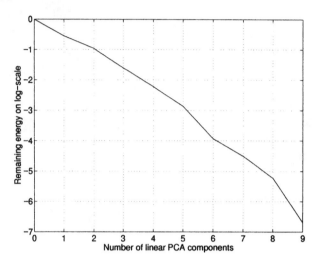

Fig. 6.7. The graph shows the remaining energy in the data as a function of the number of extracted linear PCA components. The total energy is normalised to unity (zero on the logarithmic scale). The data has been generated from five Gaussian sources but as the mapping is non-linear, the linear PCA cannot be used to find the original subspace.

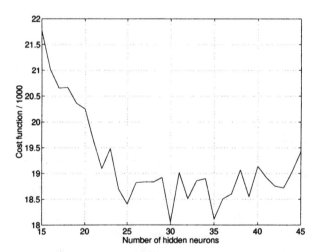

Fig. 6.8. The value of the cost function is shown as a function of the number of hidden neurons in the MLP network modelling the non-linear mapping from five sources to the observations. Ten different initialisations were tested to find the minimum value for each number of hidden neurons. The cost function exhibits a broad and somewhat noisy minimum. The smallest value for the cost function was obtained using 30 hidden neurons.

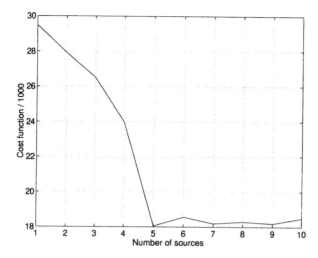

Fig. 6.9. The value of the cost function is shown as a function of the number of sources. The MLP network had 30 hidden neurons. Ten different initialisations were tested to find the minimum value for each number of sources. The cost function saturates after five sources and the deviations are due to different random initialisations of the network.

With non-linear factor analysis by MLP networks, not only the number of sources but also the number of hidden neurons in the MLP network needs to be estimated. With the Bayesian approach this is not a problem, as is shown in figures 6.8 and 6.9. The cost function exhibits a broad minimum as a function of hidden neurons and a saturating minimum as a function of sources. The reason why the cost function saturates as a function of sources is that the network is able to effectively prune away unused sources. In the case of ten sources, for instance, the network actually uses only five of them.

The pressure to prune away hidden neurons is not as big, as can be seen in figure 6.10. A reliable sign of pruning is the number of bits which the network uses to describe a variable. Recall that it was shown in section 5.4.2 that the cost function can be interpreted as the description length of the data. The description length can also be computed for each variable separately and this is shown in figure 6.10. The MLP network had seven input neurons, i.e., seven sources, and 100 hidden neurons. The upper left plot shows clearly that the network effectively uses only five of the sources and very few bits are used to describe the other two sources. This is evident also from the first layer weight matrix A on the upper right plot, which shows the average description length of the weights leaving each input neuron.

The lower plot of figure 6.10 also shows the average description length of the weight matrix A, but now the average is taken row-wise and thus tells how many bits are used to describe the weights arriving at each hidden neuron. It

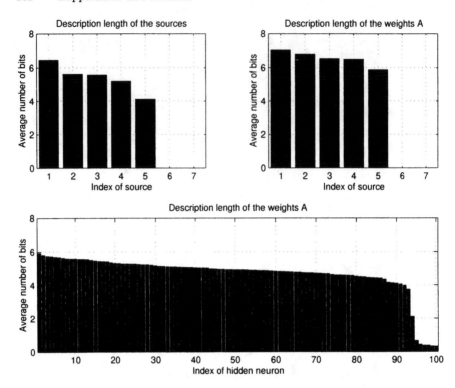

Fig. 6.10. The network is able to prune away unused parts. This can be monitored by measuring the description length of different variables. The sources and hidden neurons are sorted by decreasing description length.

appears that about six or seven hidden neurons have been pruned away, but the pruning is not as complete as in the case of sources. This is because for each source the network has to represent 1000 values, one for each observation vector, but for each hidden neuron the network only needs to represent five plus twenty (the effective number of input and output) values and there is thus much less pressure to prune away a hidden neuron.

Non-Gaussian Sources. The following experiments demonstrate the ability of the non-linear independent factor analysis algorithm to find the underlying latent variables which have generated the observations.

In these experiments, a similar scheme was used to generate data as with the Gaussian sources. A total of eight sources was used with four sub-Gaussian and four super-Gaussian sources. The generating MLP network was also larger, having 30 hidden neurons and 20 output neurons. The super-Gaussian sources were generated by taking a hyperbolic sine, $\sinh x$, from a normally distributed random variable and then normalising the variance to unity. In generating sub-Gaussian sources, inverse hyperbolic sine, $\sinh^{-1} x$, was used instead of $\sinh x$.

Several different numbers of hidden neurons were tested in order to optimise the structure of the network, but the number of sources was assumed to be known. This assumption is reasonable since it is possible to optimise the number of sources simply by minimising the cost function as the experiments with the Gaussian sources show. The network which minimised the cost function turned out to have 50 hidden neurons. The number of Gaussians in each of the mixtures modelling the distribution of each source was chosen to be three and no attempt was made to optimise this.

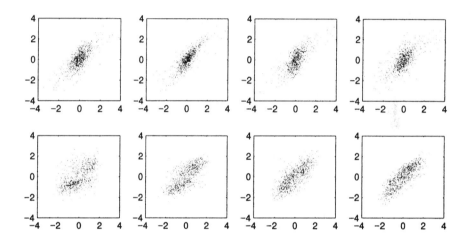

Fig. 6.11. Original sources are on the x-axis of each scatter plot and the sources estimated by a linear ICA are on the y-axis. Signal to noise ratio is 0.7 dB.

The results are depicted in figures 6.11, 6.12 and 6.13. Each figure shows eight scatter plots, each of which corresponds to one of the eight sources. The original source which was used to generate the data is on the x-axis and the estimated sources on the y-axis of each plot. Each point corresponds to one data vector. The upper plots of each figure correspond to the super-Gaussian and the lower plots to the sub-Gaussian sources. An optimal result would be a straight line which would mean that the estimated values of the sources coincide with the true values.

Figure 6.11 shows the result of a linear FastICA algorithm [4]. The linear ICA is able to retrieve the original sources with 0.7 dB signal to noise ratio. In practice, a linear method could not deduce the number of sources and the result would be even worse. The poor signal to noise ratio shows that the data really lies in a non-linear subspace.

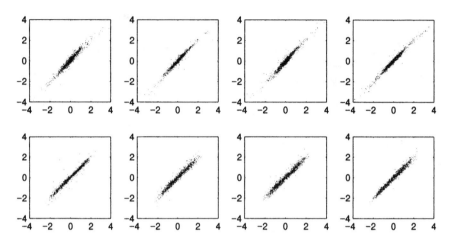

Fig. 6.12. Scatter plots of the sources after 2000 iterations of non-linear factor analysis followed by a rotation with a linear ICA. Signal to noise ratio is 13.2 dB.

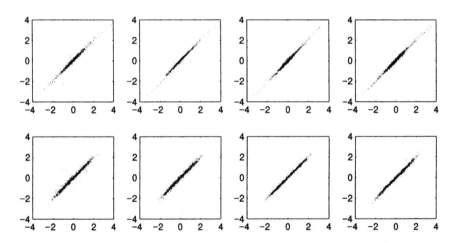

Fig. 6.13. The network in figure 6.12 has been further trained for 5500 iterations with non-linear independent factor analysis. Signal to noise ratio is 17.3 dB.

Figure 6.12 depicts the results after 2000 iterations with non-linear factor analysis followed by a rotation with a linear FastICA. Now the signal to noise ratio is 13.2 dB and the sources have clearly been retrieved. Figure 6.13 shows the final result after another 5500 iterations with a non-linear independent factor analysis algorithm. The signal to noise ratio has further improved to 17.3 dB.

It would be possible to avoid using the non-linear independent factor analysis algorithm by first running 7500 iterations with the linear factor analysis algorithm and then applying the linear ICA. The disadvantage would be that the cost function would not take into account the non-Gaussianity. The signal to noise ratio after 7500 iterations with the linear factor analysis algorithm followed by the linear ICA was 14.9 dB, which shows that taking the non-Gaussianity into account during estimation of the non-linear mapping helps the non-linear independent factor analysis algorithm to find better estimates for the sources.

6.5.4   Process Data

This data set consists of 30 time series of length 2480 measured from sensors in an industrial pulp process. An expert has preprocessed the signals by roughly compensating for time-lags of the process which originate from the finite speed of pulp flow through the process.

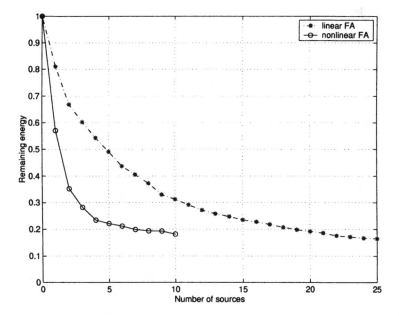

Fig. 6.14. The graph shows the remaining energy in the process data as a function of the number of extracted components in linear and non-linear factor analysis.

In order to get an idea of the dimensionality of the data, linear factor analysis was applied to the data. The result is shown in figure 6.14. The same figure also shows the results with non-linear factor analysis. It appears that the data is quite non-linear since the non-linear factor analysis is able to explain as much data with ten components as the linear factor analysis with 21 components.

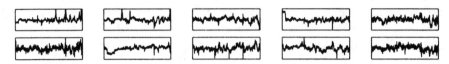

Fig. 6.15. The ten estimated sources from the industrial pulp process. Time increases from left to right.

Several different numbers of hidden neurons and sources were tested with different random initialisations with non-linear factor analysis and it turned out that the cost function was minimised for a network having ten sources and 30 hidden neurons. The same network was chosen for non-linear independent factor analysis, i.e., after 2000 iterations with linear factor analysis the sources were rotated with FastICA and each source was modelled with a mixture of three Gaussian distributions. The resulting sources are shown in figure 6.15.

Figure 6.16 shows the 30 original time series of the data set, one time series per plot, and in the same plots below the original time series are the reconstructions made by the network, i.e., the posterior means of the output of the network when the inputs were the estimated sources shown in figure 6.15. The original signals show great variability but the reconstructions are strikingly accurate. In some cases it even seems that the reconstruction is less noisy than the original signal. This is somewhat surprising since the time dependencies in the signal were not included in the model. The observation vectors could be arbitrarily shuffled and the model would still give the same result.

Initial studies are pointing to the conclusion that the estimated source signals can have meaningful physical interpretations. The results are encouraging but further studies are needed to verify the interpretations of the signals.

6.6   Comparison with Existing Methods

The idea of representing the data with a non-linear coordinate system is by no means new and several algorithms for learning the coordinate system have been proposed.

6.6.1   SOM and GTM

Self-organising maps (SOM) [5] and generative topographic mapping (GTM) [2] define a non-linear coordinate system by stating the coordinates of lattice points called model vectors. The methods are in wide use, particularly

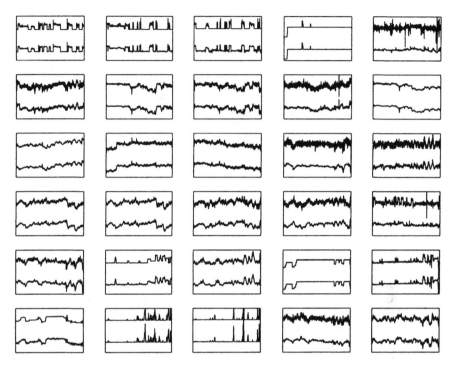

Fig. 6.16. The 30 original time series are shown on each plot above the reconstruction made from the sources shown in figure 6.15

the computationally efficient SOM, but the dimension of the latent space is normally quite small. Two-dimensional latent space is the most typical one because it allows an easy visualisation for human users.

The disadvantage of SOM and GTM is that the number of parameters required to describe the mapping from latent variables to observations grows exponentially with the dimension of the latent space. Our main motivation for using an MLP network as the non-linear mapping is that its parameterisation scales linearly with the dimension of the latent space. In this respect the mapping of an MLP network is much closer to a linear mapping which has been proven to be applicable in very high-dimensional latent spaces. SOM and GTM would probably be better models for the helical data set in section 6.5.2, but the rest of the experiments have latent spaces whose dimensions are so large that SOM or GTM models would need very many parameters.

6.6.2   Auto-Associative MLPs

Auto-associative MLP networks have been used for learning similar mappings as we have done. Both the generative model and its inversion are learned simultaneously, but separately without utilising the fact that the models are connected. This means that the learning is much slower than in this case where the inversion is defined as a gradient descent process.

Much of the work with auto-associative MLPs uses point estimates for weights and sources. As argued in the beginning of the chapter, it is then impossible to reliably choose the structure of the model and problems with over- or underlearning may be severe. Hochreiter and Schmidhuber [3] have used an MDL-based method which does estimate the distribution of the weights but has no model for the sources. It is then impossible to measure the description length of the sources.

6.6.3   Generative Learning with MLPs

MacKay and Gibbs [6] briefly report using stochastic approximation to learn a generative MLP network which they called a density network because the model defines a density of the observations. Although the results are encouraging, they do not prove the advantages of the method over SOM or GTM because the model is very simple; noise level is not estimated from the observations and the latent space had only two dimensions. The computational complexity of the method is significantly greater than in the parametric approximation of the posterior presented here, but it might be possible to combine the methods by finding an initial approximation of the posterior probability with parametric approximation and then refining it with more elaborate stochastic approximation.

In [7], a generative MLP network was optimised by gradient based learning. The cost function was the reconstruction error of the data and a point estimate was used for all the unknown variables. As argued in section 6.2, this means that it is not possible to optimise the model structure and the method is prone to overfitting.

6.7   Conclusion

6.7.1   Validity of the Approximations

The posterior pdf of all the unknown variables was approximated with a Gaussian density with diagonal covariance, which means that the variables were assumed independent given the observations. The Gaussianity assumption is not severe since the parameterisation is chosen so as to make the posterior close to Gaussian. If the hidden neurons were linear, the posterior of the sources, weights and biases would, in fact, be exactly Gaussian. Gaussian approximation therefore penalises strong non-linearities to some extent.

The posterior independence seems to be the most unrealistic assumption. It is probable that a change in one of the weights can be compensated for by changing the values of other weights and sources, which means that they have posterior dependencies.

In general, the cost function tries to make the approximation of the posterior more accurate, which means that during learning the posterior will also try to be more independent. In PCA, the mapping has a degeneracy which is used by the algorithm to do exactly this. In linear PCA the mapping is such that the sources are independent a posteriori. In the non-linear factor analysis, the dependencies of the sources are different in different parts of the latent space and it would be reasonable to model these dependencies. Computational load would not increase significantly since the Jacobian matrix computed in the algorithm can be used also for estimating the posterior interdependencies of the sources. For the sake of simplicity, the derivations were not included here.

It should be possible to do the same for the non-linear independent factor analysis, but it would probably be necessary to assume the Gaussians of each source to be independent. Otherwise the posterior approximation of $Q(M|X)$ would be computationally too intensive.

The other approximation was done when approximating the non-linearities of the hidden neurons by Taylor's series expansions. For small variances this is valid and it is therefore good to check that the variances of the inputs for the hidden neurons are not outside the range where the approximation is valid. In the computation of the gradients, some terms were neglected to discourage the network from adapting itself to areas of parameter space where the approximation is inaccurate. Experiments have proven that this seems to work. For the network which minimised the cost function in figure 6.8, for instance, the maximum variance of the input for a hidden neuron was 0.06. This maximum value is safely below the values where the approximation could become too inaccurate.

6.7.2   Initial Inversion by Auxiliary MLP

During learning, both the sources and the mapping of the network evolve. When the network is presented new data, it is necessary to find the estimates of the sources corresponding to the new data. This can be difficult using a similar update process as was used in learning because it is possible that during learning the network develops local minima which make later inversion difficult.

Experiments have shown that it is possible to learn an auxiliary MLP network which will estimate the mapping from observations to sources and can thus be used to initialise the sources, given new data. The resulting system with two MLP networks resembles an auto-associative MLP network. As was argued before, learning only the generative model is faster than learning a deeper auto-associative MLP with both the generative model and its inverse. Initial experiments have also shown that updates of the sources after the initialisation with the auxiliary MLP network lead to better estimates of the sources.

6.7.3   Future Directions

In principle, both the non-linear factor analysis and independent factor analysis can model any time-independent distribution of the observations. MLP networks are universal approximators for mappings and mixture-of-Gaussians for densities. This does not mean, however, that the models described here would be optimal for any time-independent data sets, but the Bayesian methods which were used in the derivation of the algorithms allow easy extension to more complicated models. It is also easy to use Bayesian model comparison to decide which model is most suited to the data set at hand.

An important extension would be the modelling of dependencies between consecutive sources $s(t)$ and $s(t+1)$ because many natural data sets are time series. For instance both the speech and process data sets used in the experiments clearly have strong time-dependencies.

In the Bayesian framework, treatment of missing values is simple, which opens up interesting possibilities for the non-linear models described here. A typical pattern recognition task can often be divided into unsupervised feature extraction and supervised recognition phases. Using the proposed method, the MLP network can be used for both phases. The data set for the unsupervised feature extraction would have only the raw data and the classifications would be missing. The data set for the supervised phase would include both the raw data and the desired outputs and the network. From the point of view of the method presented here, there is no need to make a clear distinction between unsupervised and supervised learning phases as any data vectors can have any combination of missing values. The network will model the joint distribution of all the observations and it is not necessary to specify which of the variables are the classifications and which are the raw data.

6.8   Acknowledgements

The authors wish to thank Xavier Giannakopoulos, M.Sc., for his help with the simulations with the non-linear independent factor analysis and Esa Alhoniemi, M.Sc., for his help with the pulp process data.

References

1. H. Attias. Independent factor analysis. *Neural Computation* **11**, (4): 803–851, 1999.
2. C.M. Bishop, M. Svensén, C.K.I. Williams. GTM: The generative topographic mapping. *Neural Computation* **10**, (1): 215–234, 1998.
3. S. Hochreiter and J. Schmidhuber. LOCOCODE performs non-linear ICA without knowing the number of sources. *Proceedings of the ICA '99,* 149–154, 1999.
4. A. Hyvärinen and E. Oja. A fast fixed-point algorithm for independent component analysis. *Neural Computation* **9**,(7): 1483–1492, 1997.

5. T. Kohonen: *Self-Organizing Maps*. (Springer, New York, 1995.)
6. D.J.C. MacKay and M.N. Gibbs. Density networks. In J.Kay, editor, *Proceedings of Society for General Microbiology Edinburgh Meeting*, 1997.
7. J.-H. Oh and H.S. Seung. Learning generative models with the up-propagation algorithm. In M. I. Jordan, M. J. Kearns, S. A. Solla, editors, *Advances in Neural Information Processing Systems 10*,605–611, MIT Press, Cambridge, MA, 1998.

7 Ensemble Learning for Blind Image Separation and Deconvolution

James Miskin and David J. C. MacKay

7.1 Introduction

In this chapter, ensemble learning is applied to the problem of blind source separation and deconvolution of images. It is assumed that the observed images were constructed by mixing a set of images (consisting of independent, identically distributed pixels), convolving the mixtures with unknown blurring filters and then adding Gaussian noise.

Ensemble learning is used to approximate the intractable posterior distribution over the unknown images and unknown filters by a simpler separable distribution. The mixture of Laplacians used for the source prior respects the positivity of the image and favours sparse images. The model is trained by minimising the Kullback-Leibler divergence between the true posterior distribution and the approximating ensemble.

Unlike maximum-likelihood methods, increasing the number of hidden images does not lead to overfitting the data and so the number of hidden images in the observed data can be inferred.

The results show that the algorithm is able to deconvolve and separate the images and correctly identify the number of hidden images.

Previous work on blind source deconvolution has focused mainly on the problem of deconvolving sound samples. It is assumed that the observed sound samples are temporally convolved versions of the true source samples. Blind deconvolution algorithms have fallen into two types, those where the inverse of the convolution filter is learnt [1,3] and those where the aim is to learn the filter itself [1].

Two problems become apparent when applying these ideas to the problem of deconvolving images. First, in many real data sets (for instance, the images generated by telescopes observing the sky or the power spectrum from a nuclear magnetic resonance (NMR) spectrometer) the pixel values correspond to intensities. So the pixel values must be positive. The standard blind separation approaches of assuming that the sources are distributed as $\frac{1}{\cosh}$ [3] or mixtures of Gaussians [2] lose this positivity of the source images. Deconvolution without a positivity constraint leads to reconstructed images that have areas of negative intensity corresponding to energy being sucked out of the detector. Lack of a positivity constraint explains why an optimal linear filter is suboptimal for deconvolution. The derivation of the optimal linear filter assumes that the source image consists of independent identically distributed Gaussian pixels and so does not force positivity [4,8].

Another problem is that we know that the convolution filters must also be positive (in the case of observing the sky, the convolution filter corresponds to blurring due to an imperfect telescope). Most algorithms that learn the inverses of the filters do not take this into account, particularly since positivity is only true of the convolution filter and not of the deconvolution filter. Algorithms that learn the convolution filter also have problems if they assume that the filter can be inverted exactly, which is not necessarily the case since there may be zeros in the power spectrum of the convolution filter which lead to poles in the power spectrum of the deconvolution filter. Poles in the power spectrum of the deconvolution filter would not be a problem if the observed image were noise-free and there were no numerical errors in the calculation but in real problems the inverse is ill-conditioned.

It is possible to use MCMC sampling to solve blind inverse problems by sampling from the true posterior density for the latent variables [7]. Sampling methods have the disadvantage that predictions can only be made by storing a set of samples or by repeating the sampling process whenever samples are needed.

In this chapter we apply ensemble learning (as introduced in chapter 5) to the problem of blind source separation and deconvolution. We apply ensemble learning so as to fit the ideal posterior density by an approximation satisfying the positivity constraints. The result of the training is a distribution (not a set of samples from the distribution) and so further inferences can be made by evaluating expectations using this approximation. Ensemble learning has previously been applied to the blind source separation problem where it was used to separate a sample of speech into independent components [5].

We apply the image separation algorithm to the problem of finding a representation for a sub-set of the MNIST handwritten digit database. We show that the digit images can be represented by a smaller set of localised images representing parts of the digits. Factorisation of data using a positive constraint on latent variables has previously been used to find a parts-based representation for images of faces and for text passages [6].

7.2   Separation of Images

We will consider separating a linear mixture of hidden source images. The set of N observed images (each of which is I by J pixels) is assumed to be given by

$$
\begin{aligned}
y_{ij,n} &= \sum_{m=1}^{M} w_{nm} x_{ij,m} + \nu_{ij,n} \\
&= \hat{y}_{ij,n} + \nu_{ij,n}
\end{aligned}
\tag{7.1}
$$

where w is an N by M matrix, x is the set of M hidden images (each I by J pixels) and ν is some zero mean Gaussian noise.

The priors over the latent variables (w and x) must respect the positivity constraint. The prior for the matrix elements is a Laplacian:

$$p(w_{nm}) = \begin{cases} \beta_w \exp(-\beta_w w_{nm}) & w_{nm} \geq 0 \\ 0 & w_{nm} < 0 \end{cases} \qquad (7.2)$$

where β_w is a scale parameter with the scale invariant prior

$$p(\ln \beta_w) = 1 \qquad (7.3)$$

The prior for the source pixels is a mixture of Laplacians:

$$p(x_{ij,n}) = \begin{cases} \sum_{\alpha=1}^{N_\alpha} \pi_\alpha \frac{1}{b_\alpha} \exp\left(-\frac{x_{ij,m}}{b_\alpha}\right) & x_{ij,m} \geq 0 \\ 0 & x_{ij,m} < 0 \end{cases} \qquad (7.4)$$

where the hyper-priors are

$$p(\ln b_\alpha) = 1 \qquad (7.5)$$

$$p(\{\pi_\alpha\}) \propto \delta\left(\sum_{\alpha=1}^{N_\alpha} \pi_\alpha - 1\right) \prod_{\alpha=1}^{N_\alpha} \pi_\alpha . \qquad (7.6)$$

Figure 7.1 shows how a prior of this form favours sparse images. It should be noted that the prior does not include any prior knowledge we may have about spatial structure.

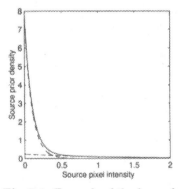 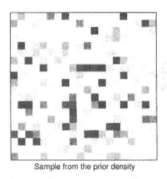

Fig. 7.1. Example of the form of the source pixel prior density. The sum of Laplacians can have a sharp peak at zero intensity and a long tail. Consequently the prior favours sparse source images.

If we assume that we observe the mixtures with additive Gaussian noise, then the likelihood for the observed pixels is

$$p(\{y\} | \{x\}, \{w\}, \beta_\sigma) = \prod_{ij,n} \mathcal{G}\left(y_{ij,n}; \hat{y}_{ij,n}, \beta_\sigma^{-1}\right) \qquad (7.7)$$

where $\mathcal{G}(a; b, c)$ is a Gaussian distribution over a with mean b and variance c. β_σ is the inverse variance of the Gaussian noise and is assigned the hyperprior

$$p(\ln \beta_\sigma) = 1 \qquad (7.8)$$

We now define the model $\mathcal{H}$ to be the variables M, $\{b_\alpha\}$ and $\{\pi_\alpha\}$ and Θ to be the latent variables $\{x\}$, $\{w\}$, β_σ and β_w. Using Bayes theorem, the posterior density over the latent variables is

$$p(\Theta \,|\{y\}, \mathcal{H}) = \frac{p(\{y\}\,|\Theta, \mathcal{H})\, p(\Theta)}{p(\{y\}, \mathcal{H})} \qquad (7.9)$$

The process of making inferences involves finding expectations under this probability density (typically expectations of the latent variables themselves), which is analytically intractable. As shown in chapter 5, we can approximate the true posterior by a more tractable distribution, $q(\{x\}, \{w\}, \beta_\sigma, \beta_w)$, for which the expectations are tractable. We can do this by minimising the cost function

$$
\begin{aligned}
C_{\mathrm{KL}} &= D\left(q\left(\Theta\right)\|p\left(\Theta\,|\{y\}, \mathcal{H}\right)\right) - \ln p\left(\{y\}\,|\mathcal{H}\right) - \ln p\left(\{b_\alpha\}\right) - \ln p\left(\{\pi_\alpha\}\right) \\
&= \int q(\Theta) \ln \frac{q(\Theta)}{p\left(\Theta, \{y\}\,|\mathcal{H}\right)} d\Theta - \ln p\left(\{b_\alpha\}\right) - \ln p\left(\{\pi_\alpha\}\right) \\
&= \int \left[\ln \frac{q(\{w\})}{p(\{w\})} + \ln \frac{q(\{x\})}{p(\{x\})} + \ln \frac{q(\beta_\sigma)}{p(\beta_\sigma)} \right. \\
&\qquad \left. + \ln \frac{q(\beta_w)}{p(\beta_w)} - \ln p\left(\{y\}\,|\Theta, \mathcal{H}\right) \right] d\Theta \\
&\qquad - \ln p\left(\{b_\alpha\}\right) - \ln p\left(\{\pi_\alpha\}\right) \qquad (7.10)
\end{aligned}
$$

It should be noted that because of the product form of the true posterior density, the cost function can be written as a sum of simpler terms.

7.2.1   Learning the Ensemble

In order to simplify the posterior density, we choose to use a separable distribution of the form

$$q(\Theta) = \prod_{ijm} q(x_{ij,m}) \times \prod_{nm} q(w_{nm}) \times q(\beta_\sigma)\, q(\beta_w) \qquad (7.11)$$

We will not assume a specific form for these distributions, instead we will find the set of functions that optimises the cost function (subject to the separable form and the constraint that each distribution is normalised).

We can update each distribution in turn, using current estimates for all of the other distributions. To illustrate this, we can consider performing all of the integrations in the cost function with the exception of the integration over β_w

$$
C_{\text{KL}} = \int q\left(\beta_w\right) \left[\ln q\left(\beta_w\right) - \sum_{nm} \left(\ln \beta_w - \beta_w \left\langle w_{nm}\right\rangle\right) + \ln \beta_w\right] d\beta_w
$$

(7.12)

where we have dropped all terms that are independent of β_w and $\langle.\rangle$ denotes the expectation under the approximating ensemble. We now need to minimise this cost function with respect to the distribution $q\left(\beta_w\right)$, subject to the constraint that $q\left(\beta_w\right)$ is normalised.

$$
\frac{\partial C_{\text{KL}}}{\partial q\left(\beta_w\right)} = \ln q\left(\beta_w\right) - \sum_{nm} \left(\ln \beta_w - \beta_w \left\langle w_{nm}\right\rangle\right) + \ln \beta_w + 1 + \lambda_w
$$

(7.13)

where λ_w is a Lagrange multiplier. Setting this derivative to zero, we find that the optimum distribution for β_w is

$$
\ln q\left(\beta_w\right) = \sum_{nm} \left[\ln \beta_w - \beta_w \left\langle w_{nm}\right\rangle\right] - \ln \beta_w - 1 - \lambda_w
$$

(7.14)

Therefore the optimal distribution is

$$
q\left(\beta_w\right) = \Gamma\left(\beta_w; \sum_{nm} \left\langle w_{nm}\right\rangle, NM\right)
$$

(7.15)

where the Γ distribution is

$$
\Gamma\left(a; b, c\right) = \frac{1}{\Gamma\left(c\right)} b^c a^{(c-1)} \exp\left(-ab\right)
$$

(7.16)

Similarly we find that the optimal distribution for β_σ is

$$
q\left(\beta_\sigma\right) = \Gamma\left(\beta_\sigma; \frac{1}{2}\sum_{ijn} \left\langle \left(y_{ij,n} - \hat{y}_{ij,n}\right)^2\right\rangle, \frac{IJN}{2}\right)
$$

(7.17)

For the remaining parameters the optimal distributions are

$$
q\left(w_{nm}\right) = \frac{1}{Z_{nm}^{(w)}} p\left(w_{nm}\right) \exp\left(-\frac{1}{2}w_{nm}^{(2)}\left(w_{nm} - w_{nm}^{(1)}\right)^2\right)
$$

(7.18)

$$
q\left(x_{ij,m}\right) = \frac{1}{Z_{ij,m}^{(x)}} p\left(x_{ij,m}\right) \exp\left(-\frac{1}{2}x_{ij,m}^{(2)}\left(x_{ij,m} - x_{ij,m}^{(1)}\right)^2\right)
$$

(7.19)

where $\left\{ Z_{nm}^{(w)} \right\}$ and $\left\{ Z_{ij,m}^{(x)} \right\}$ are the sets of normalising constants and we have defined

$$w_{nm}^{(2)} = \frac{1}{\langle \beta_\sigma \rangle} \sum_{ij} \langle x_{ij,m}^2 \rangle \tag{7.20}$$

$$w_{nm}^{(1)} w_{nm}^{(2)} = \langle \beta_\sigma \rangle \sum_{ij} \langle x_{ij,m} \rangle \left(y_{ij,n} - \sum_{m' \neq m} \langle w_{nm'} \rangle \langle x_{ij,m'} \rangle \right) \tag{7.21}$$

$$x_{ij,m}^{(2)} = \frac{1}{\langle \beta_\sigma \rangle} \sum_{n} \langle w_{nm}^2 \rangle \tag{7.22}$$

$$x_{ij,m}^{(1)} x_{ij,m}^{(2)} = \langle \beta_\sigma \rangle \sum_{n} \langle w_{nm} \rangle \left(y_{ij,n} - \sum_{m' \neq m} \langle w_{nm'} \rangle \langle x_{ij,m'} \rangle \right) \tag{7.23}$$

The optimal distributions for w are products of Laplacians and Gaussians, so the optimal distributions are rectified Gaussians (i.e. $q(w_{nm})$ is Gaussian for $w_{nm} \geq 0$ and zero otherwise). Similarly the optimal distributions for x are mixtures of rectified Gaussians. When evaluating the updates for the distributions, it is necessary to evaluate the expectations of the form $\langle w_{nm} \rangle$, $\langle w_{nm}^2 \rangle$, etc. These can be evaluated using error functions.

The distributions can be trained by repeatedly updating each one in turn. But it is important to note that, while we have chosen the approximate ensemble such that samples from the distributions are independent, the parameters of the distributions are correlated and so optimisation by successive updating of each distribution can be slow to converge.

We can update all of the distributions in parallel by noting that the ensemble can be parameterised by the vector

$$\boldsymbol{\theta} = \left(\left\{ x^{(1)} \right\}, \left\{ \log x^{(2)} \right\}, \left\{ w^{(1)} \right\}, \left\{ \log w^{(2)} \right\}, \log a_w, \log a_\sigma \right) \tag{7.24}$$

where

$$a_w = \sum_{nm} \langle w_{nm} \rangle \tag{7.25}$$

$$a_\sigma = \frac{1}{2} \sum_{ijn} \left\langle (y_{ij,n} - \hat{y}_{ij,n})^2 \right\rangle \tag{7.26}$$

The current estimate of the ensemble can be parameterised by $\boldsymbol{\theta}^{(\tau)}$. We can then define the vector $\boldsymbol{\theta}^{(\text{opt})}$ to be the ensemble formed from the optimal distributions according to equations 7.15, 7.17 and 7.20 – 7.23. A small step along the vector from $\boldsymbol{\theta}^{(\tau)}$ to $\boldsymbol{\theta}^{(\text{opt})}$ will reduce the cost function. Therefore the new ensemble can be defined to be the minimum along the vector from $\boldsymbol{\theta}^{(\tau)}$ to $\boldsymbol{\theta}^{(\text{opt})}$.

If the distributions are independent, a single line minimisation will result in convergence to the optimum distribution (since in this case $\boldsymbol{\theta}^{(\tau+1)} =$

$\theta^{(\text{opt})}$). Alternatively if the distributions are not independent, successive line minimisations will result in convergence to the optimum ensemble.

7.2.2   Learning the Model

We would also like to be able to infer the parameters of the prior on the source pixels. We can do this by noting that the terms in the cost function relating to the prior on the source pixels are

$$
C_{\text{KL}} = -\ln p\left(\{b_\alpha\}\right) - \ln p\left(\{\pi_\alpha\}\right) - \sum_{ij,m} \langle \ln p\left(x_{ij,m}\right)\rangle
$$

$$
= -\ln p\left(\{b_\alpha\}\right) - \ln p\left(\{\pi_\alpha\}\right)
$$
$$
- \sum_{ij,m} \int q\left(x_{ij,m}\right) \ln \sum_\alpha \pi_\alpha \frac{1}{b_\alpha} \exp\left(-\frac{x_{ij,m}}{b_\alpha}\right) dx_{ij,m} \qquad (7.27)
$$

If the current parameters are $\left\{b_\alpha^{(\tau)}\right\}$ and $\left\{\pi_\alpha^{(\tau)}\right\}$ and the updated parameter values are $\left\{b_\alpha^{(\tau+1)}\right\}$ and $\left\{\pi_\alpha^{(\tau+1)}\right\}$, then by application of Jensen's inequality and discarding constants, a bound on the cost function can be obtained

$$
C_{\text{KL}} \leq -\sum_{ij,m,\alpha} \int f_{\alpha,i,j,m}\left(x_{ij,m}\right) \ln\left[\pi_\alpha^{(\tau+1)} \frac{1}{b_\alpha^{(\tau+1)}} \exp\left(-\frac{x_{ij,m}}{b_\alpha^{(\tau+1)}}\right)\right] dx_{ij,m}
$$
$$
- \ln p\left(\left\{b_\alpha^{(\tau+1)}\right\}\right) - \ln p\left(\left\{\pi_\alpha^{(\tau+1)}\right\}\right) \qquad (7.28)
$$

where

$$
f_{\alpha,i,j,m}\left(x_{ij,m}\right) = \frac{1}{Z_{ij,m}^{(x)}} \pi_\alpha^{(\tau)} \frac{1}{b_\alpha^{(\tau)}} \exp\left(-\frac{x_{ij,m}}{b_\alpha^{(\tau)}}\right)
$$
$$
\times \exp\left(-\frac{1}{2}x_{ij,m}^{(2)}\left(x_{ij,m} - x_{ij,m}^{(1)}\right)^2\right) \qquad (7.29)
$$

The bound on the cost function can be optimised by setting the new parameters for the prior to

$$
\pi_\alpha^{(\tau+1)} = \frac{1 + \sum_{ijm}\left[\int f_{\alpha,i,j,m}\left(x_{ij,m}\right) dx_{ij,m}\right]}{IJM + N_\alpha} \qquad (7.30)
$$

$$
b_\alpha^{(\tau+1)} = \frac{\sum_{ijm}\left[\int f_{\alpha,i,j,m}\left(x_{ij,m}\right) x_{ij,m} dx_{ij,m}\right]}{1 + \sum_{ijm}\left[\int f_{\alpha,i,j,m}\left(x_{ij,m}\right) dx_{ij,m}\right]} \qquad (7.31)
$$

7.2.3   Example

Figure 7.2 shows the results of separating a mixture of three grey-scale Dilbert images. The images were mixed with a random positive matrix and Gaussian noise was added. The three columns of the figure show the true hidden images, the noisy observations and the ensemble average for the reconstructed images. Three of the reconstructed images match the hidden images. The other two images do not contribute to the mixture. The elements in the w matrix corresponding to those images are set to approximately zero. If we look at equation 7.22, the $x^{(2)}$ parameters tend to zero as the elements of w tend to zero and so the posterior density for all of the pixels in the blank images matches the prior density.

We can see that the posterior tends to the prior by looking at figure 7.3 where the KL divergence between the posterior and prior source pixel densities for each reconstructed image is plotted as a function of iteration. It can be seen that the divergence tends to zero for two of the images which means that the posterior and the prior are the same densities.

It might be useful to infer the number of source images that contribute to the observation. In maximum likelihood methods, increasing the number of sources cannot decrease the likelihood since the extra source images will model the noise in the observations. Therefore the number of source images will be inferred to be at least as large as the number of observed images.

The correct way to perform the inference of the number of sources is to perform model selection, where each model corresponds to a different number of hidden images. The cost function gives us a bound on the evidence for a model,

$$\ln p\left(\{y\} | \mathcal{H}\right) \geq -C_{\mathrm{KL}} \tag{7.32}$$

Therefore we can use Bayes theorem to evaluate the posterior probability of a given model using

$$p\left(\mathcal{H} | \{y\}\right) = \frac{p\left(\{y\} | \mathcal{H}\right) p\left(\mathcal{H}\right)}{p\left(\{y\}\right)} \tag{7.33}$$

If we choose a flat prior over the number of hidden images, the model that maximises the posterior distribution is the model that maximises $p\left(\{y\} | \mathcal{H}\right)$. We could assume that this is the same as the model that maximises the bound on $p\left(\{y\} | \mathcal{H}\right)$ and so the model to choose is the model that minimises C_{KL}.

It may be too time-consuming to train multiple models, one for each possible number of images. A simpler method would be to remove source images from the model that do not contribute to the observations. This will reduce the cost function since we know that the KL divergence between the posterior and prior densities for the source pixels and for the mixing matrix elements must be greater than zero. Practically, we remove a source image if the KL divergence between the posterior and prior densities drops to less than 10^{-3} per pixel.

Figure 7.8 shows how the histograms of intensity compare for the hidden images, the observed images and the recovered images. It should be noted that the recovered images are much more sparse than the observed images.

Fig. 7.2. Separation of a mixture of three images from a set of five observed images. The left-hand column shows the true hidden images. The centre column shows the noisy mixtures of the hidden images. The right-hand column shows the reconstructed source images. Three of the reconstructed images match the true images; the remaining two images are uniform as their approximate posterior density, $q(x_{ij,m})$, is equivalent to the prior density. [Dilbert image Copyright©1997 United Feature Syndicate, Inc., used with permission.]

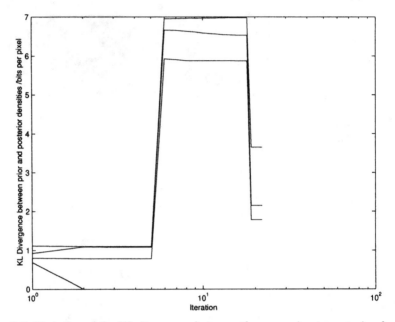

Fig. 7.3. Variation of the KL divergence between the approximate posterior density and the prior density for each of the reconstructed images. There are three stages to training. During the first stage, the source prior and $q\,(\beta_\sigma)$ are not trained. During the second stage, $q\,(\beta_\sigma)$ is trained so the approximate posterior distributions become sharper and the KL divergence increases. During the final stage, the source prior is updated so that it better fits the approximate posterior and the KL divergence drops.

7.2.4   Parts-Based Image Decomposition

Positive constraints on latent pixels have previously been used to find a non-negative factorisation of a set of face data [6]. In that case, it was found that the reconstructed images corresponded to a parts-based decomposition of the face data into localised features.

We can consider trying to find images in a set of natural images by using the EL blind separation algorithm. Figure 7.4 shows the first 16 examples of handwritten "3"s in the MNIST data set. Figure 7.5 shows the first 16 PCA components generated from the first 256 "3"s in the MNIST data set. These components represent the highest variance components of the data set, but it is not obvious visually what the set of components represents.

The PCA components do not respect the known positivity of the images (the digits range from white to black or zero ink to lots of ink). Therefore when the PCA components are added together there is an interaction between positive and negative regions in different components to give the positive digit images.

Fig. 7.4. The first 16 "3"s in the MNIST handwritten data set. The digits are stored as 28x28 pixel grey-scale images. Each digit has been preprocessed to centre it in the image and to deskew it.

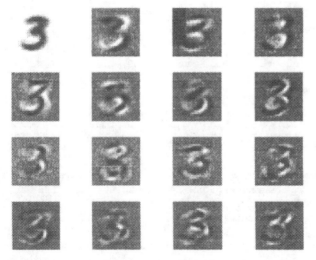

Fig. 7.5. The first 16 PCA components from the first 256 "3"s in the MNIST data set.

We can consider enforcing positivity of the latent images. Applying the ensemble learning algorithm to the set of 256 "3"s (assuming that there are 64 hidden images) leads to the decomposition shown in figure 7.6. Instead of the images being based on corrections to a prototype "3" (as in the PCA case) the reconstructed images are all localised and take the form of different shapes of curves, tails, etc. Figure 7.7 shows the reconstructions of the digits in figure 7.4 using the learnt hidden images. Therefore a parts based decomposition can give a good representation of the data set.

The parts-based representation could be used for image compression (by storing images in terms of the parts required to construct them) or as a method of image recognition (by training a set of models of different digits, "1"s, "2"s, etc, a classifier could be made by evaluating the posterior probability of each model for each trial digit).

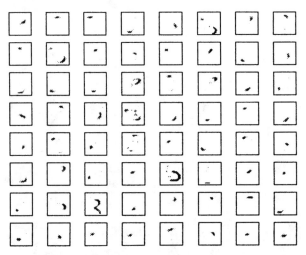

Fig. 7.6. The hidden images learnt when the 256 "3"s are assumed to be made from a linear combination of 64 non-negative latent images. The learnt images are localised (unlike the PCA components) and each represents a part of a "3" with different images representing the shapes of curves in different positions.

7.3 Deconvolution of Images

We can extend the model to include localised blurring of the images. The model for blurring could be used to model the point-spread function for a telescope, the line width in NMR experiments or motion blur. The observed

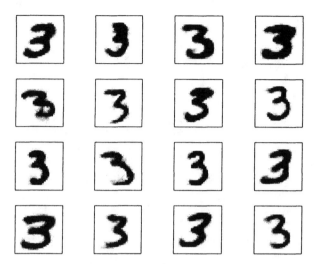

Fig. 7.7. Reconstructions of the true digits using the 64 non-negative images. Here we can see that the parts-based representation is able to model a variety of shapes of "3" and we can see that a set of 64 latent images is able to represent the data set of 256 handwritten "3"s.

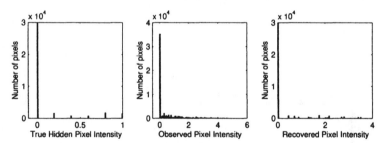

Fig. 7.8. Histograms of the intensities of the pixels in the images. The first plot shows the true images and we can see that the images are quantised grey-scale images. The second plot shows the observed images; as there is a mixing of the quantised levels, the observations are no longer quantised. The third plot shows the reconstructed images. The reconstructed images do not match the true images exactly, because the ICA model has an invariance with respect to rescaling each source image, but the reconstructed images are more sparse than the observed images.

images are now defined by

$$
\begin{aligned}
y_{ij,n} &= \sum_{m=1}^{M} \sum_{k=-K}^{K} \sum_{l=-K}^{K} w_{nm} e_{kl,n} x_{i-k,j-l,m} + \nu_{ij,n} \\
&= \hat{y}_{ij,n} + \nu_{ij,n}
\end{aligned}
\tag{7.34}
$$

where w, x, ν and y have the same definitions as the previous model and e is a set of localised convolution filters (one for each image) which extend from $-K$ to K in each dimension. In evaluating this sum, it is assumed that $x_{ij,m}$ is zero outside the defined extent of the image.

The priors for this model are the same as for the previous model with the addition of a prior for the convolution filters. The prior is similar to the prior for the mixing parameters and respects the positivity of the filter.

$$
p\left(e_{kl,n}\right) = \begin{cases} \beta_e \exp\left(-\beta_e e_{kl,n}\right) & e_{kl,n} \geq 0 \\ 0 & e_{kl,n} < 0 \end{cases}
\tag{7.35}
$$

where β_e is a scale parameter with the scale invariant prior

$$
p\left(\ln \beta_e\right) = 1
\tag{7.36}
$$

We can now approximate the true posterior by the separable distribution

$$
q\left(\Theta\right) = \prod_{ijm} q\left(x_{ij,m}\right) \times \prod_{nm} q\left(w_{nm}\right) \times \prod_{kln} q\left(e_{kl,n}\right) \times q\left(\beta_\sigma\right) q\left(\beta_w\right) q\left(\beta_e\right)
\tag{7.37}
$$

Again we do not assume a specific form for the distributions in $q\left(\Theta\right)$. If we find the optimal distributions we obtain

$$
q\left(\beta_w\right) = \Gamma\left(\beta_w; \sum_{nm} \langle w_{nm}\rangle, NM\right)
\tag{7.38}
$$

$$
q\left(\beta_e\right) = \Gamma\left(\beta_e; \sum_{kln} \langle e_{kl,n}\rangle, (2K+1)^2 N\right)
\tag{7.39}
$$

$$
q\left(\beta_\sigma\right) = \Gamma\left(\beta_\sigma; \frac{1}{2}\sum_{ijn} \left\langle (y_{ij,n} - \hat{y}_{ij,n})^2\right\rangle, \frac{IJN}{2}\right)
\tag{7.40}
$$

$$
q\left(w_{nm}\right) = \frac{1}{Z_{nm}^{(w)}} p\left(w_{nm}\right) \exp\left(-\frac{1}{2} w_{nm}^{(2)} \left(w_{nm} - w_{nm}^{(1)}\right)^2\right)
\tag{7.41}
$$

$$
q\left(e_{kl,n}\right) = \frac{1}{Z_{kl,n}^{(e)}} p\left(e_{kl,n}\right) \exp\left(-\frac{1}{2} e_{kl,n}^{(2)} \left(e_{kl,n} - e_{kl,n}^{(1)}\right)^2\right)
\tag{7.42}
$$

$$
q\left(x_{ij,m}\right) = \frac{1}{Z_{ij,m}^{(x)}} p\left(x_{ij,m}\right) \exp\left(-\frac{1}{2} x_{ij,m}^{(2)} \left(x_{ij,m} - x_{ij,m}^{(1)}\right)^2\right)
\tag{7.43}
$$

where we have defined

$$w_{nm}^{(2)} = \frac{1}{\langle\beta_\sigma\rangle} \sum_{ij} \left[\left(\langle e_{kl,n}\rangle \langle x_{i-k,j-l,m}\rangle \right)^2 \right.$$

$$\left. + \sum_{kl} \left(\langle e_{kl,n}^2\rangle \langle x_{i-k,j-l,m}^2\rangle - \langle e_{kl,n}\rangle^2 \langle x_{i-k,j-l,m}\rangle^2 \right) \right] \qquad (7.44)$$

$$w_{nm}^{(1)} w_{nm}^{(2)} = \langle\beta_\sigma\rangle \sum_{ij} \left[\sum_{kl} \langle e_{kl,n}\rangle \langle x_{i-k,j-l,m}\rangle\, y_{ij,n} \right.$$

$$\left. - \sum_{k_1 l_1 k_2 l_2} \sum_{m'\neq m} \langle w_{nm'} e_{k_1 l_1,n} e_{k_2 l_2,n} x_{i-k_1,j-l_1,m} x_{i-k_2,j-l_2,m'}\rangle \right]$$

$$(7.45)$$

$$e_{kl,n}^{(2)} = \frac{1}{\langle\beta_\sigma\rangle} \sum_{ij} \left[\left(\sum_m \langle w_{nm}\rangle \langle x_{i-k,j-l,m}\rangle \right)^2 \right.$$

$$\left. + \sum_m \left(\langle w_{nm}^2\rangle \langle x_{i-k,j-l,m}^2\rangle - \langle w_{nm}\rangle^2 \langle x_{i-k,j-l,m}\rangle^2 \right) \right] \qquad (7.46)$$

$$e_{kl,n}^{(1)} e_{kl,n}^{(2)} = \frac{1}{\langle\beta_\sigma\rangle} \sum_{ij} \left[\sum_m \langle w_{nm}\rangle \langle x_{i-k,j-l,m}\rangle\, y_{ij,n} \right.$$

$$\left. - \sum_{m_1 m_2} \sum_{k_2 l_2\neq kl} \langle w_{nm_1} w_{nm_2} e_{k_2 l_2,n} x_{i-k,j-l,m_1} x_{i-k_2,j-l_2,m_2}\rangle \right] \quad (7.47)$$

$$x_{ij,m}^{(2)} = \frac{1}{\langle\beta_\sigma\rangle} \sum_{i'j'n} \langle w_{nm}^2\rangle \langle e_{i'-i,j'-j,n}^2\rangle \qquad (7.48)$$

$$x_{ij,m}^{(1)} x_{ij,m}^{(2)} = \langle\beta_\sigma\rangle \sum_{i_2 j_2 n} \left[\langle w_{nm}\rangle \langle e_{i_2-i,j_2-j,n}\rangle\, y_{i_2 j_2,n} \right.$$

$$\left. - \sum_{m_2 k_2 l_2\neq m i_2-ij_2-j} \langle w_{nm} w_{nm_2} e_{i_2-ij_2-j,n} e_{k_2 l_2,n} x_{i_2-k,j_2-l,m_2}\rangle \right]$$

$$(7.49)$$

The posterior distributions can be trained iteratively by performing repeated line minimisations as for the separation of images model.

Figure 7.9 shows the results of using the ensemble learning algorithm to reconstruct the hidden image and the blurring filter from a single observed image. In each case the reconstructed filter matches the true filter. The reconstructed images match the true hidden images.

Figure 7.10 shows the results of using the Ensemble Learning algorithm to reconstruct the hidden images and the blurring filters from a set of blurred images (the images are the same as those used in figure 7.2, but with added blurring). In each case the reconstructed filter matches the true filter and the reconstructed image also matches the true hidden image.

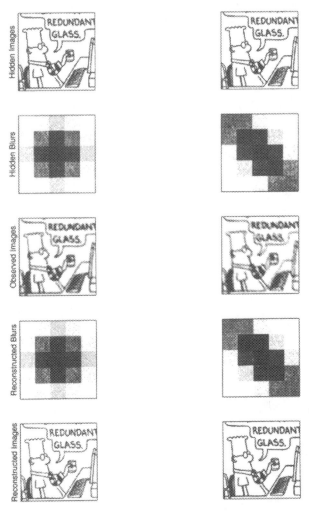

Fig. 7.9. The deconvolution of two blurred images. In each test the same image was blurred by a different filter. The reconstructed filters match the true filters. The reconstructed images are close to the hidden images. [Dilbert image Copyright©1997 United Feature Syndicate, Inc., used with permission.]

Figure 7.11 shows how the histograms of intensity compare for the hidden images, the observed images and the recovered images. As with the pure

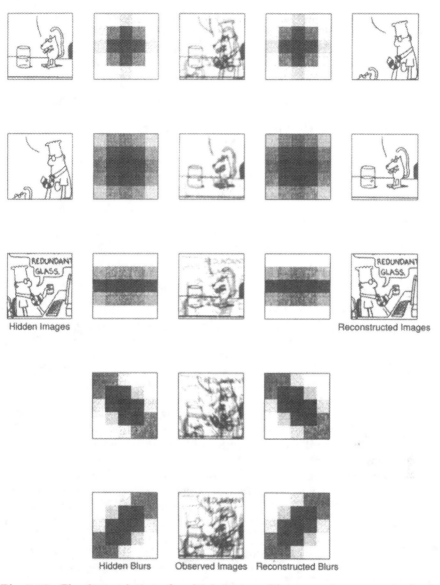

Fig. 7.10. The deconvolution of multiple images. The source images were mixed and then blurred by a set of localised blur filters. The reconstructed images match the source images, showing that the correct mixing matrix and blurring filters were learnt. [Dilbert image Copyright©1997 United Feature Syndicate, Inc., used with permission.]

mixing case, the observed images are much less sparse than the true hidden images. The choice of a sparse prior for the images helps to force a set of sparse reconstructed source images to be found.

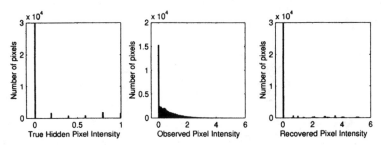

Fig. 7.11. Histograms of the intensities of the pixels in the multiple blurred images. The first plot shows the true images, here we can see that the images are quantised grey-scale images. The second plot shows the plot for the observed images; as there is a mixing of the quantised levels, the observations are no longer quantised. The third plot shows the histogram for the reconstructed source images.

7.4   Conclusion

Freeform ensemble learning allows for tractable solutions to blind inverse problems. Approximating the true posterior by a more tractable separable distribution means that the blind inverse problem can be reduced to a function minimisation problem. Consequently the inverse need not be performed by resorting to an MCMC sampler.

A side effect of using a separable approximating posterior distribution is that correlations between the latent variables in the true posterior distribution are lost. On the other hand, the fitting process uses distributions over possible values for all of the parameters, unlike maximum likelihood methods which find a point estimate for the latent variables and consequently can suffer from overfitting to the data.

Use of ensemble learning allows the number of hidden images to be inferred by minimising the cost function (or equivalently maximising the bound on the evidence) with respect to the number of hidden images.

The results show that the algorithm is able to deconvolve and separate noisy mixtures of images. The results also show that the algorithm can be used to obtain a parts-based representation of images.

For hidden images that have an intrinsic correlation, the images could be modelled by a set of independent pixels (as in the model described above) convolved with another unknown blurring filter. The extra filter could be learnt in a similar way to the filter in this model and may improve separation of images that have intrinsic correlations.

7.5   Acknowledgements

The authors would like to thank Harri Lappalainen for useful discussions and Yann LeCun for the use of the MNIST data set.

References

1. H. Attias. Blind source separation and deconvolution: The dynamic component analysis algorithm. *Neural Computation* **10**: 1373–1424, 1998.
2. H. Attias. Independent factor analysis. *Neural Computation* **11**: 803–851, 1998.
3. A. J. Bell and T. J. Sejnowski. An information maximisation approach to blind separation and blind deconvolution. *Neural Computation.* **7**: 1129–1159, 1995.
4. S. F. Gull and G. J. Daniell. Image reconstruction from incomplete and noisy data. *Nature.* **272**: 686–690, 1978.
5. H. Lappalainen. Ensemble learning for independent component analysis. *Proceedings of the First International Workshop on Independent Component Analysis and Blind Signal Separation*, 1998.
6. D D. Lee and H. S. Seung. Learning the parts of objects by non-negative matrix factorization. *Nature* **401**: 788–791, 1999.
7. M. F. Ochs, R. S. Stoyanova, F. Arias-Mendoza, and T. R. Brown. A new method for spectral decomposition using a bilinear bayesian approach. *Journal of Magnetic Resonance.* **137**: 161–176, 1999.
8. J E. Tansley, M J. Oldfield, and D J. C. Mackay: Neural network image deconvolution. In *Maximum Entropy and Bayesian Methods.* G. R. Heidbreder, editor, 319–325, Kluwer Academic Publishers, 1996.

Part IV

Data Analysis and Applications

8 Multi-Class Independent Component Analysis (MUCICA) for Rank-Deficient Distributions

Francesco Palmieri and Alessandra Budillon

8.1 Introduction

Standard linear Independent Component Analysis (ICA) has recently been extended to provide signal reconstruction for a multi-class mixture model [6,7,9]. Such an approach connects classical EM estimation to the ICA framework. Unfortunately, computational problems arise when the class densities, that underly the observed data are degenerate or are poorly conditioned. This appears to be very likely in many applications. In this chapter we approach the problem by assuming that the samples from each class are confined to different linear subspaces. The criterion is based on estimating an on-line membership function for each data point which combines a measure of the the likelihood and the distance from the subspaces. The independent components are then searched within each subspace. We present results of the algorithm on synthetic distributions with various degrees of degeneracy. Our results are promising for feature extraction applications.

One of the most effective ways of modeling vector data for unsupervised pattern classification or coding is to assume that the observations are the result of picking randomly out of a fixed set of different distributions. Gaussian mixture models are also very popular because the EM algorithm allows efficient maximum likelihood estimation of the model parameters [5]. The identification system, often reformulated as "mixture of experts", estimates the most likely current class from a computed set of a posteriori probabilities evaluated using the estimated distribution parameters. The mixed densities can also be non-Gaussian and assumed to belong to a predefined class such as, for example, the exponential family. The application of the EM algorithm is also rather straightforward, in this more general case, if the distribution that composes the mixture is non-degenerate. The identification of the mixture densities and their parameters can be thought of as a process of decomposition in two levels: one at the density level and one at the level of the coordinates of each density into marginal independent components. Such an idea has been proposed in [6,7] and by us in [9] independently in extending to this multi-class scenario the standard ICA algorithms [1,3,4]. The idea consists of running in parallel more ICA searches with the adaptation of each conditioned by the class membership of the observed data points. The results are also quite interesting because most of the results that have appeared so far in the literature on ICA can be extended to such a more general framework. It is very likely, as it has been observed in many applications, that the class densities that underly observed data, such as images or one-dimensional sequences, may be degenerate or poorly conditioned. A number of observations confirm this fact as PCA analysis and multiple PCA analysis [10], when conditioned by any kind of class discrimination, reveal class densities which are either rank-deficient or show very concentrated eigenvalue distributions. This means that, in the applications, very likely class data are confined to linear subspaces. The EM algorithm, or modifications of it with the ICA generalization [2] also in their typical formulation, needs to be modified to handle degenerate distributions. This becomes immediately evident when we have to compute the a posteriori class probability for a point that does not belong to any of the subspaces identified by the algorithm up to that point. Making the EM algorithm more robust is quite challenging in such cases and, in our experience, it is very easy to lose control of the algorithm in trying to overcome overflows and indeterminacies. In this chapter we assume that the generative model of the mixture densities is degenerate and propose approaches to invert it. In section 8.2, we review the one-class scenario by differentiating among three possible partitionings of the reconstructing block. In 8.3 we propose the rank-deficient multi-class identification and in 8.4 we report a set of simulations on synthetic data set. Some conclusions follow in section 8.5.

8.2   The Rank-Deficient One Class Problem

Figure. 8.1 (a) shows the generative model of an N-dimensional vector x. We assume that the components of x are a linear combination, with an $N \times M$ mixing matrix A, of $M \leq N$ sources assumed to be mutually independent, with zero mean and unitary variances, i.e. $x = As$, $\mathrm{E}[s] = o$, $\mathrm{E}[ss^{\mathrm{T}}] = I$. The source densities are $\{f_{s_i}(\xi)\}_{i=1}^{M}$, with $f_{s_i}(\xi) = \dot{\phi}_i(\xi)$, where $\{\phi_{s_i}(\xi)\}_{i=1}^{M}$ are the cumulative distribution functions. In the figure we also indicated a block ϕ^{-1} that in the model is supposed to shape the densities of s from M independent random variables $u = [u_1, \ldots, u_M]^{\mathrm{T}}$, uniformly distributed in $[0, 1]$. The purpose of the block B is to invert the system giving at the M-dimensional output y a copy of the source vector s.

This is the classical blind source separation (BSS) problem in the general under-complete case, in which more observations than sources are available. Source reconstruction can usually be achieved except for a permutation and a sign change. Also the scale must be arbitrary and this is why we have taken unit variance sources. BSS has been successfully achieved if B is such that:

$$BA = AS\Pi, \tag{8.1}$$

where B is an $M \times N$ matrix, A is an arbitrary scale diagonal matrix, Π is a permutation matrix and $S = \mathrm{diag}(\pm 1, \ldots, \pm 1)$ is a diagonal sign matrix. In

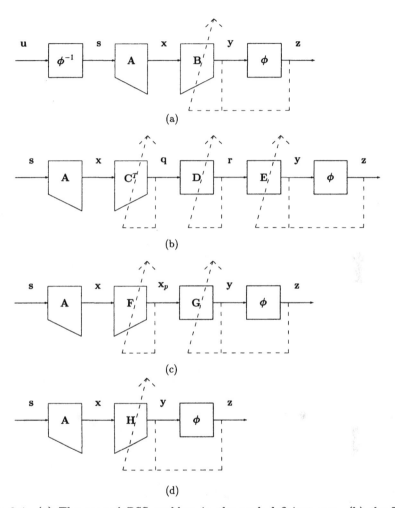

Fig. 8.1. (a) The general BSS problem in the rank-deficient case, (b) the BSS problem where the block B is split into three autonomous blocks, (c) the BSS problem where the block B is split into two autonomous blocks, (d) the BSS problem where the block B consists of only one block.

other words B, apart from the sign, must be a permuted and scaled version of the pseudoinverse of A. This is easily visualized using the singular value decomposition of A and B^T :

$$\begin{cases} A & = U_A \Lambda_A V_A^T \\ B^T & = U_B \Lambda_B V_B^T \end{cases} \qquad (8.2)$$

where U_A is the $N \times M$ matrix containing in its columns the first M left eigenvectors of A, Λ_A is diagonal and contains the M singular values of A and V_A is $M \times M$ and is the unitary matrix containing the M right eigenvectors of A. Similarly for B^T. Clearly

$$BA = V_B \Lambda_B U_B^T U_A \Lambda_A V_A^T \qquad (8.3)$$

and if

$$\begin{cases} U_B = U_A \\ \Lambda_B = \Lambda_A^{-1} \\ V_B = V_A \end{cases} \qquad (8.4)$$

we have $BA = I$. More generally if

$$\begin{cases} U_B = U_A \\ \Lambda_B = \Lambda_A^{-1} \\ V_B = \Lambda S \Pi V_A \end{cases} \qquad (8.5)$$

we have $BA = \Lambda S \Pi$. This means that

$$B^T = U_A \Lambda_A^{-1} V_A^T \Pi^T S \Lambda = A^{\#T} \Pi^T S \Lambda \qquad (8.6)$$

where $A^{\#}$ is the pseudoinverse of A. In solving a multi-class problem, in which more separating paths must be learned at the same time, it would be desirable to identify a compact approach to the one-class problem. In fact covering out parallel learning on the various branches could become cumbersome when too many variables are changing simultaneously. In the following we review three possible partitionings of the problem as background to the derivation of the multi-class learning equations.

8.2.1   Method I: Three Blocks

A first classical approach to finding B is to decompose the problem into three sub-blocks, with $B = EDC^T$, as shown in Figure 8.1 (b), as if the structure had to reproduce somehow the singular value decomposition of $A^{\#}$. Each block in the cascade can be trained by itself and can be brought to converge once the previous ones have converged. In particular, the first block C^T has to project the data onto the linear subspace on which they lie. In fact the space of $x = As$ will contain data only on the hyperplane determined by the left eigenvectors of A. Such a subspace can be easily identified since from a sequence of observed x, the correlation matrix $E[xx^T] = AE[ss^T]A^T = AA^T$

can be estimated. The eigenvectors of AA^T are (by definition) the left eigenvectors of A, i.e. $AA^T = U_A \Lambda_A V_A^T V_A \Lambda_A U_A^T = U_A \Lambda_A^2 U_A^T$. Furthermore, if the singular values in $\Lambda_A^2 = \text{diag}(\lambda_1^2, \ldots, \lambda_M^2)$ are kept in decreasing order of magnitude, i.e. $\lambda_1^2 \leq \lambda_2^2 \ldots \leq \lambda_M^2$, the determination of U_A is unique and the first block is a principal component analyzer. The choice $C = U_A$ gives $E[qq^T] = U_A^T E[xx^T] U_A = \Lambda_A^2$, i.e. C is a decorrelating block. The next diagonal block has to "sphere" the data, i.e. $E[rr^T] = I$, D can simply be chosen as

$$D = E[qq^T]^{-1/2} = (\Lambda_A^2)^{-1/2} = \text{diag}(|\lambda_{A_1}|^{-1}, \ldots, |\lambda_{A_M}|^{-1}) = S\Lambda_A^{-1}$$

(8.7)

To complete the pseudoinverse of A, it is necessary to add a new block. This is an orthogonal $M \times M$ matrix E made of M right eigenvectors of A, perhaps scaled and permutated. The M right eigenvectors of A are the first M non null eigenvectors of $A^T A$ and must be learned by imposing high order independence on the components of y. Note that all the second order information has been removed ($E[rr^T] = I$) and that the correlation matrix remains unitary only for unitary matrices E:

$$E[yy^T] = EE[rr^T]E^T = I$$

One of the well-known ICA standard algorithms [1,3,4] can be used to evaluate E.

8.2.2 Method II : Two Blocks

Figure 8.1 (c) shows a two blocks approach with $B = GF$. In particular, the first block F, as previously, has to project the data onto the linear subspace on which they lie. It is sufficient that F spans the first M eigenvectors space, i.e. it can be chosen equal to RU_A^T, where R is any rotation matrix. This means that F can be searched minimizing the mean squared reconstruction error, i.e. $F = \underset{F}{\text{argmin}} E[\|x - F^T Fx\|^2]$. It has been proven that this cost function provides an orthonormal basis of the M-dimensional principal component subspace of the correlation matrix $E[xx^T]$. Then the square matrix G should force the independence of the projections $x_p = Fx$ onto the M-dimensional subspace. Since we do not sphere the data we cannot assume that the input has identity correlation matrix. The singular value decomposition also in this case provides a clear description of the process. If

$$\begin{cases} F = RU_A^T \\ G = \Lambda S \Pi V_A \Lambda_A^{-1} R^T \end{cases}$$

(8.8)

we have

$$\begin{aligned} BA &= GFA \\ &= \Lambda S \Pi V_A \Lambda_A^{-1} R^T R U_A^T U_A \Lambda_A V_A^T \\ &= \Lambda S \Pi \end{aligned}$$

(8.9)

One of the well-known ICA algorithms [1,3,4] can be used to find G, just as for E.

8.2.3   Method III : One Block

Finally, the last scheme (d) of figure 8.1 consists of only one block, $B = H$. It should project the data onto the linear subspace on which they lie and it should make these projections as independent as possible. To recover the independent sources, H should be a scaled and permuted version of $A^\#$:

$$H = \Lambda S \Pi V_A \Lambda_A^{-1} U_A^{\mathrm{T}} \tag{8.10}$$

To obtain a unique updating rule for H we consider a composite cost function

$$C(H) = \mathrm{K}_r[\mathrm{E}[\Psi(y)y^{\mathrm{T}} - I]] + \lambda \mathrm{E}[\|x - H^\# H x\|^2] = C_1(H) + \lambda C_2(H) \tag{8.11}$$

where K_r is the Kronecker norm of the argument, defined as:

$$\mathrm{K}_r[\mathrm{E}[\Psi(y)y^{\mathrm{T}} - I]] = \sum_{i=1}^{M} \sum_{j=1}^{M} (\mathrm{E}[\psi_i(y_i)y_j] - \delta_{ij})^2 \tag{8.12}$$

and Ψ is defined as

$$\begin{aligned} \Psi &= (\psi_1, \dots, \psi_M)^{\mathrm{T}} \\ &= \left(-\frac{\dot{\phi}_1(y_1)}{\phi_1(y_1)}, \dots, -\frac{\dot{\phi}_M(y_M)}{\phi_M(y_M)} \right)^{\mathrm{T}} \end{aligned} \tag{8.13}$$

With the first cost function, C_1, we force decorrelation between the projections y and their non-linear versions $\Psi(y)$. This objective involving non linear versions of y should search for projections which are as statistically independent as possible. With the second cost function, C_2, we search for the best linear reconstruction, in the mean square error sense, by means of the projections Hx. The parameter λ represents the reciprocal weight between the two cost functions that must be chosen empirically by cross-validation. To derive the learning algorithm, we evaluate the derivatives of the two cost functions with respect to the parameter matrix H:

$$\nabla_H C = \nabla_H C_1 + \lambda \nabla_H C_2 \tag{8.14}$$

Using standard techniques from matrix differential calculus [8], for the first term we have

$$\begin{aligned} d_H C_1 &= d_H \mathrm{K}_r[\mathrm{E}[\Psi(y)y^{\mathrm{T}} - I]] \\ &= d_H(e^{\mathrm{T}}\mathrm{E}[\Psi(y)y^{\mathrm{T}} - I] \odot \mathrm{E}[\Psi(y)y^{\mathrm{T}} - I]e) \\ &= 2e^{\mathrm{T}}\mathrm{E}[\Psi(y)y^{\mathrm{T}} - I] \odot d_H \mathrm{E}[\Psi(y)y^{\mathrm{T}} - I]e \\ &= 2\mathrm{tr}[\mathrm{E}[xy^{\mathrm{T}}\mathrm{E}[\Psi(y)y^{\mathrm{T}} - I]^{\mathrm{T}}\mathrm{diag}\dot{\Psi}(y))d_H H + \\ &\quad \mathrm{E}[x\Psi(y)^{\mathrm{T}}]\mathrm{E}[\Psi(y)y^{\mathrm{T}} - I]d_H H] \end{aligned} \tag{8.15}$$

where $e = (1, \ldots, 1)^{\mathrm{T}}$, $\odot$ is the Hadamard product and tr is the trace, which gives:

$$
\begin{aligned}
\nabla_H C_1 = {} & 2\mathrm{E}[\mathrm{diag}\dot{\Psi}(y)\mathrm{E}[\Psi(y)y^{\mathrm{T}} - I]yx^{\mathrm{T}}] \\
& + \mathrm{E}[\Psi(y)y^{\mathrm{T}} - I]^{\mathrm{T}}\mathrm{E}[\Psi(y)x^{\mathrm{T}}]
\end{aligned}
\tag{8.16}
$$

For the second term we have

$$
\begin{aligned}
d_H C_2 &= d_H \mathrm{E}[\|x - H^{\#}Hx\|^2] \\
&= 2\mathrm{E}[(x - H^{\#}Hx)^{\mathrm{T}}d_H(x - H^{\#}Hx)] \\
&= -2\mathrm{tr}[\mathrm{E}[(I - H^{\#}H)((I - H^{\#}H)xx^{\mathrm{T}}H^{\#}d_H H + \\
& \qquad x^{\mathrm{T}}x(I - H^{\#}H)^{\mathrm{T}}H^{\#}d_H H)]]
\end{aligned}
\tag{8.17}
$$

which gives:

$$
\begin{aligned}
\nabla_H C_2 = -2\mathrm{E}[& H^{\#\mathrm{T}}(xx^{\mathrm{T}}(I - H^{\#}H) + \\
& (xx^{\mathrm{T}}(I - H^{\#}H))^{\mathrm{T}})(I - H^{\#}H)]
\end{aligned}
\tag{8.18}
$$

A gradient descent algorithm to update the parameter H would be:

$$
\Delta H = -\eta(\nabla_H C_1 + \lambda \nabla_H C_2).
\tag{8.19}
$$

The algorithm can be used in batch by computing empirical means from the data in a stochastic approximation using instantaneous values. We found experimentally that the batch approach provides better results.

8.3   The Rank-Deficient Multi-Class Problem

Assume now that the random vector $x \in \mathcal{R}^N$ is distributed according to a parametric mixture density model $f_x(x, \Theta) = \sum_{i=1}^{C} \pi_i f_x(x|i, \theta_i)$, where C is the known number of classes, $\{\pi\}_{i=1,\ldots,C}$ are the a priori probabilities (supposed to be known or set to $1/C$) and $\Theta = (\theta_1, \ldots, \theta_C)$ are the unknown mixture parameters. As depicted in figure 8.2, we train in parallel C structures of the type shown in figure 8.1 (b)-(d), each one performing an affine transformation $y_i = B_i(x - \mu_i)$ with parameters (B_i, μ_i) and a sigmoidal function $\phi_i = (\phi_{i1}, \ldots, \phi_{iM})^{\mathrm{T}}$ obtained from a set of cumulative distribution functions as in the one-class problem. We indicate with $y_i = (y_{i1}, \ldots, y_{iM})^{\mathrm{T}}$ and $z_i = (z_{i1}, \ldots, z_{iM})^{\mathrm{T}}$, the inputs and the outputs to the sigmoids respectively.

We could apply the results obtained in the one-class scenario if we were able to estimate the class membership for each datum. We have already treated this situation [9] when the matrices B are square using the observation that if the independence had been accomplished one could substitute

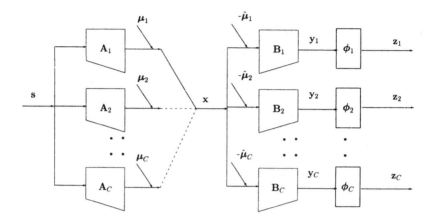

Fig. 8.2. The ICA structure for mixtures in the case $M \leq N$.

$f_{\boldsymbol{y}_i}(\boldsymbol{y}_i) = \prod_{j=1}^{M} f_{s_{ij}}(y_{ij})$. This is equivalent to applying the EM algorithm. Unfortunately if $\boldsymbol{x}$ does not belong to any of the hyperplanes defined by the current set of $\{\boldsymbol{B}_i\}$, the posterior probability is not defined. Therefore to attribute to a generic observation, $\boldsymbol{x}$, a given class, we adopt the following membership function:

$$MF(i, \Theta, \boldsymbol{x}) = \frac{\exp(-d(\boldsymbol{x} - \boldsymbol{\mu}_i, \boldsymbol{B}_i))}{\sum_{j=1}^{C} \exp(-d(\boldsymbol{x} - \boldsymbol{\mu}_j, \boldsymbol{B}_j))} f_{\boldsymbol{y}_i}(\boldsymbol{B}_i(\boldsymbol{x} - \boldsymbol{\mu}_i)) \qquad (8.20)$$

where $\Theta = (\boldsymbol{B}_1, \boldsymbol{\mu}_1, \ldots, \boldsymbol{B}_C, \boldsymbol{\mu}_C)$ and $d(\boldsymbol{x} - \boldsymbol{\mu}_i, \boldsymbol{B}_i) = \|\boldsymbol{x} - \boldsymbol{\mu}_i - \boldsymbol{B}_i^{\#}\boldsymbol{B}_i(\boldsymbol{x} - \boldsymbol{\mu}_i)\|$ is the square distance of $\boldsymbol{x} - \boldsymbol{\mu}_i$ from the hyperplane generated by $\boldsymbol{B}_i$. The first term represents a relative measure of the distance of the data from the hyperplanes and the second term is the likelihood on that hyperplane. Given the current estimation $\hat{\Theta}$, we associate $\boldsymbol{x}$ to the class k^{th} if

$$k = \operatorname*{argmax}_{1 \leq i \leq C} MF(i, \hat{\Theta}, \boldsymbol{x}) \qquad (8.21)$$

Note that if $\boldsymbol{x} - \boldsymbol{\mu}_i$ belongs to the hyperplane identified by $\boldsymbol{B}_i$, $d = 0$. If $\boldsymbol{x}$ is equidistant from two or more hyperplanes, the likelihood will have a major weight in the decision and if the likelihood is similar for different classes, the class associated to the closest hyperplane will be chosen. Figure 8.3 shows the hyperplane generated by the row vectors $\boldsymbol{b}_1^{\mathrm{T}}$ and $\boldsymbol{b}_2^{\mathrm{T}}$ of a 2×3 matrix $\boldsymbol{B}$, a generic sample $\boldsymbol{x}$, its projection $\boldsymbol{x}_p$ onto the hyperplane and the distance from the hyperplane.

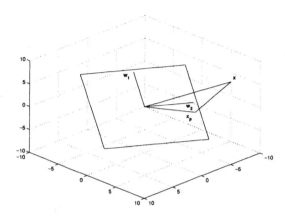

Fig. 8.3. A hyperplane generated by a matrix 2×3, a generic sample x, its projection x_p and the distance $d(x, B)$.

The algorithm proceeds in parallel on the various classes as follows:

1. Initialize $\hat{\Theta}(0) = \{B_i(0), \mu_i(0)\}_{i=1}^{C}$ to random values.

2. Present an input x and forward propagate it (with fixed weights), computing $\dot{\phi}_{ij}(y_{ij})$ and $\psi_{ij}(y_{ij})$ $\forall i = 1, \ldots, C$ and $\forall j = 1, \ldots, M$. Then compute $MF(i, \hat{\Theta}(n), x)$ from equation 8.20 $\forall i$, select the class k^{th} as in equation 8.21 and store x in a set $\mathcal{X}_k$.

3. Repeat step (2) until the whole training set has been presented.

4. $\forall$ class i update $\hat{\Theta}_i(n+1)$ from the data of its set $\mathcal{X}_i$ according to the rule:

$$\begin{cases} B_i(n+1) = B_i(n) - \eta(\nabla_{B_i} C_{1i} + \lambda \nabla_{B_i} C_{2i}) \\ \mu_i(n+1) = \frac{\sum_{\{x \in \mathcal{X}_i\}}}{card(\mathcal{X}_i)} \end{cases} \quad (8.22)$$

(Alternatively use different updating rules for the three block or two block cases.)

5. Go back to step (2) until convergence. Remember that $C_{1i} = K_r[E[\Psi_i(y_i)y_i^T - I]]|_{y_i = B_i(x - \mu_i)}, x \in \mathcal{X}_i$ and $C_{2i} = E[\|(x - \mu_i) - B_i^{\#} B_i(x - \mu_i)\|^2]|_{x \in \mathcal{X}_i}$.

8.4 Simulations

We report a set of simulations on synthetic data to verify the behaviour of the various algorithms. A first set of simulations aims to compare the behaviour of the one, two and three block approaches to learning a degenerate distribution. Figure 8.4 shows a set of the 3-dimensional observed data (o), artificially generated as a linear combination, with a 3×2 mixing matrix, of two logistic sources.

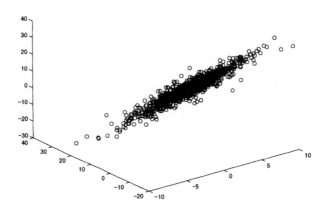

Fig. 8.4. Data from the training set in the one-class case.

On this data, we have tried the approaches outlined in section 8.2. Learning has been performed in batch on 1000 samples and figures 8.5, 8.6 and 8.7 show the evaluation of B which should converge to the identity matrix or a permutation of it.

The sigmoidal function has been chosen to force the outputs to have unit variance as the sources. For the three-block case we have used the algorithm with $\eta = 10^{-4}$; for the two-block cases, $\eta_1 = \eta_2 = 10^{-4}$; for the one-block case we have used the gradient algorithm of section 8.3 with $\eta = 10^{-4}$ and $\lambda = 10^{-1}$. The one-block case, even though it is much more compact in its formulation, gives much smaller convergence. We have not attempted to optimize λ and η, but we have found that generally it is more difficult to bring the algorithm to convergence because the cost function may be poorly conditioned. Figure 8.8 shows 1000 points generated by the model learned with the one-block algorithm. The two densities are very similar and the other algorithms show similar performance.

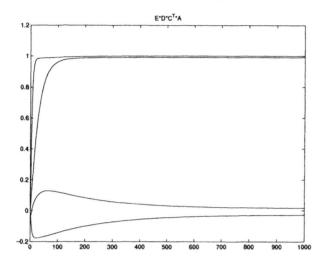

Fig. 8.5. The matrix $\boldsymbol{E} * \boldsymbol{D} * \boldsymbol{C}^{\mathrm{T}} * \boldsymbol{A}$, with respect to the epochs using the three-block approach in the one-class case.

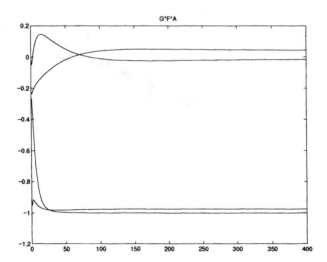

Fig. 8.6. The matrix $\boldsymbol{G} * \boldsymbol{F} * \boldsymbol{A}$, with respect to the epochs using the two-block approach in the one-class case.

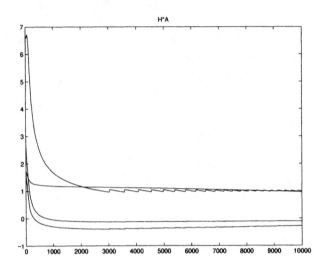

Fig. 8.7. The matrix $H * A$, with respect to the epochs using the one-block approach in the one-class case.

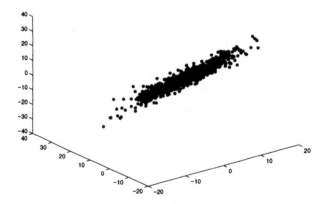

Fig. 8.8. Data generated from the model learned by the one-block algorithm in the one-class case.

For the multi-class problem, we have generated 10,000 points coming from four 3×2 mixing matrices having logistic independent sources. Figures 8.9 and 8.10 show respectively the training set and a synthetic distribution obtained with the parameters after learning.

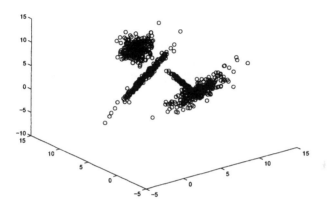

Fig. 8.9. Data from the training set in the four-class case (o)

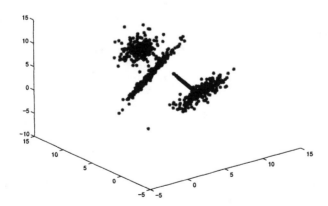

Fig. 8.10. Data generated from the model learned by the one-block algorithm in the four-class case.

Figures 8.11 and 8.12 show evaluation of the two cost functions, C_1i and C_2i for the four classes.

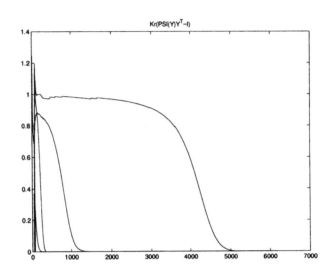

Fig. 8.11. The cost function $C_1(\boldsymbol{H}) = \mathrm{K}_r[\mathrm{E}[\boldsymbol{\Psi}(\boldsymbol{y})\boldsymbol{y}^{\mathrm{T}} - \boldsymbol{I}]]$ with respect to the epochs using the one-block approach in the four-class case.

The parameters chosen were $\eta = 10^{-4}$ and $\lambda = 10^{-1}$. Note how the second cost function converges faster than the first one, 300 epochs against 6000 epochs.

Other algorithms based on the cascade of two and three blocks show similar performance.

8.5   Conclusion

In this chapter we address the problem of separating multi-class sources which have either rank-deficient distributions or show very concentrated eigenvalues. We assume from the beginning that the mixture densities are degenerate and we update in parallel the parameters of their separating blocks. The class membership of each point is based on a distance measure from the hyperplanes and on the likelihood of each hyperplane. Moreover, we have proposed a compact approach, with only one linear block, to perform independent component analysis even if more observations than sources are available (usually approached with two blocks: a principal components analyzer followed by a standard independent components analyzer). The motivation was to obtain

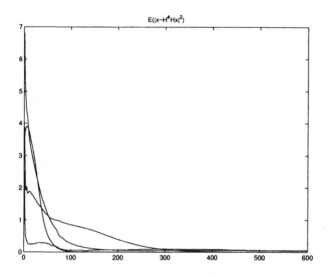

Fig. 8.12. The cost function $C_2(\boldsymbol{H}) = \mathrm{E}[\|\boldsymbol{x} - \boldsymbol{H}^{\#}\boldsymbol{H}\boldsymbol{x}\|^2]$ with respect to the epochs using the one-block approach in the four-class case.

more compact formulation for the multi-class scenario. Unfortunately this approach seems to converge more slowly than the others and this probably depends on the cost function behaviour.

Open problems, such as the choice of the number of classes, the rank of each class, the choice of the step-size η and of the weight λ are still under investigation.

References

1. S. Amari, A. Cichocki, and H. Yang. A new learning algorithm for blind signal separation. In D. Touretzky, M. Mozer, and M. Hasselmo, editors, *Advances in Neural Information Processing Systems*, **8**, 757–763, MIT Press, Cambridge MA, 1996.
2. H. Attias. Independent factor analysis. *Neural Computation*, **11**(5):803–852, 1999.
3. A. J. Bell and T. J. Sejnowski: A non-linear information maximization algorithm that performs blind signal separation. In G. Tesauro, D. S. Touretzky and T. K. Leen, editors,*Advances in Neural Information Processing Systems 7.*, 467-474, MIT Press, Cambridge, MA, 1995.
4. J-F. Cardoso. Infomax and maximum likelihood for blind separation. *IEEE Sig. Proc. Letters*, **4**(4):112–114, 1997.

5. A. P. Dempster, N. M. Laird, and D. B. Rubin. Maximum likelihood from incomplete data via the EM algorithm. *Journal of the Royal Statistical Society-B*, **39**:1–38, 1977.

6. J. Karhunen, and S. Malaroiu. Local independent component analysis using clustering. *International Workshop on Independent Component Analysis, ICA'99, Aussois, France, Jan. 1999.*

7. T-W. Lee, M. S. Lewicki and T. S. Sejnowski. ICA Mixture Models For Unsupervised Classification And Automatic Context Switching. *Proc. International Workshop on Independent Component Analysis, ICA'99, Aussois, France, Jan. 1999.*

8. J. R. Magnus and H. Neudecker. *Matrix Differential Calculus with Applications in Statistics and Econometrics.* (John Wiley & Sons, 1988.

9. F. Palmieri, D. Mattera and A. Budillon. Independent Component Analysis for Mixture Densities. *Proc. of European Symposium on Artificial Neural Networks, ESANN'99, Bruge, Belgium, Apr. 1999.*

10. M. E. Tipping and C. M. Bishop. Mixtures of Probabilistic Principal Component Analyzers. *Neural Computation*, **11**, 443-482, 1999.

9 Blind Separation of Noisy Image Mixtures

Lars Kai Hansen

9.1 Introduction

Reconstruction of statistically independent source signals from linear mixtures is relevant to many signal processing contexts [1,3,6,11,22]. Considered a generalization of principal component analysis, the problem is often referred to as independent component analysis (ICA) [9].

The source separation problem has a likelihood formulation, see e.g., [2,23,12,11]. The likelihood formulation is attractive for several reasons. First, it requires a quantitative specification of the source signal prior distribution that often remains implicit in other separation schemes. The prior distribution of the source signals can take many forms and *factorizes* in the source index expressing the fact that we look for *independent* sources. Secondly, the likelihood approach allows for direct adaptation of the plethora of powerful schemes for parameter optimization, regularization, and evaluation of supervised learning algorithms. Finally, for the case of linear mixtures without noise, the likelihood approach is equivalent to another popular approach based on information maximization [1,3,9].

The source separation problem can be analyzed under the assumption that the sources are either independent samples, i.e., "white" sequences, or assuming a more general temporal structure. The case of *autocorrelated* sequences was studied by Molgedey and Schuster [18]. They proposed a source separation scheme based on assumed non-vanishing temporal autocorrelation functions of the independent source sequences evaluated at a specific *time lag*. Their analysis was developed for source signals mixed by square, non-singular matrices. Attias and Schreiner derived a likelihood based algorithm for separation of general sequences with a frequency domain implementation [2]. Belouchrani and Cardoso [2] presented a general likelihood approach allowing for *additive noise* and for non-square mixing matrices. They applied the method to separation of sources taking discrete values [2] and estimated the mixing matrix using an Estimate - Maximize (EM) approach with both a determistic and a stochastic formulation. Moulines et al. used the EM approach for separation of autocorrelated sequences in presence of noise and explored a family of flexible source signal priors based on Gaussian Mixtures [19]. The difficult problem of noisy, overcomplete source models (more sources than mixture signals) was recently analyzed by Lewicki and Sejnowski within the likelihood framework [14].

In this chapter, we study the likelihood approach focusing on the conceptually simpler "white" source case. Our aim is to analyze short natural image sequences as found e.g., in the context of neuroimaging [9]. Hence, we address noisy mixtures. We also want to approach high-dimensional mixtures where the number of sources is much smaller than the number of sensors (pixels). An additional objective is to find models that *generalize*, i.e., models that perform well on test data.

We give a maximum posterior estimate for the sources which interestingly turns out to be non-linear in the observed signal. The specific model investigated here is a special case of the general framework proposed by Belouchrani and Cardoso [2], however we formulate the parameter estimation problem in terms of the Boltzmann learning rule, which allows for a particular transparent derivation of the mixing matrix estimate.

9.2   The Likelihood

Let the observed mixture signals be denoted X, a matrix of size $D \times N$, where D is the number of sensor channels and N is the number of observations (time points). The noisy mixing model takes the form,

$$X = AS + \Gamma \tag{9.1}$$

where S is the source signal matrix (size $M \times N$, M is the number of sources), A is the $D \times M$ mixing matrix, while Γ is a matrix of noise signals.

The properties of the source signals are introduced by a parametrized prior distribution $P(S|\psi)$. The source distribution factorizes in the source index expressing the fact that the sources are independent. In the instantaneous model analyzed by Bell and Sejnowski [3] and by MacKay [11], the distribution of S also factorizes in time, viz., assuming "white" sources.

The likelihood of the parameters of the noise distribution, the parameters of the source distribution and of the mixing matrix is then given by,

$$L(A, \theta, \psi) = P(X|A, \theta, \psi) = \int dS P(X - AS|\theta) P(S|\psi) \tag{9.2}$$

where $P(.|\theta)$ is the parameterised noise distribution. Note that the source signals play the role of "hidden" or "latent" variables. In the noise-free case the distribution $P(.|\theta)$ collapses to a delta function and the integral can be carried out by coordinate transformation [11].

For clarity we will assume that the noise can be modeled by i.i.d. Gaussian variables with variance $\theta = \sigma^2$,

$$P(\Gamma|\sigma^2) = \frac{1}{(2\pi\sigma^2)^{PN/2}} \exp\left(-\frac{1}{2\sigma^2}\mathrm{Tr}_t\Gamma^\top\Gamma\right). \tag{9.3}$$

with the matrix trace over time points in the sample ($t = 1, ..., N$). Finally, we will use the parameter-free "sigmoidal" cumulative model studied in [3,11],

$$P(S) = \frac{1}{\pi^{NM}} \exp\left(-\sum_{t,m} \log\cosh S\right) \tag{9.4}$$

the sum extending over all sources ($m = 1, ...M$) and times ($t = 1, ..., N$).

9.3 Estimation of Sources for the Case of Known Parameters

Let us first address the problem of estimating the sources if the model parameters are known, i.e., for given A, σ^2. We use Bayes formula $P(S|X) \propto P(X|S)P(S)$ to obtain the posterior distribution of the sources:

$$P(S|X, A, \sigma^2) \propto$$
$$\exp\left(-\frac{1}{2\sigma^2}\mathrm{Tr}_t((X - AS)^\top(X - AS)) - \sum_{t,m}\log\cosh S\right) \tag{9.5}$$

The *maximum a posteriori* source estimate is found by maximizing this expression w.r.t. S, providing the following non-linear equation to solve for the estimate $\widehat{S}$,

$$-A^\top A\widehat{S} + A^\top X - \sigma^2 \tanh\widehat{S} = 0 \tag{9.6}$$

There are two problems with equation 13.10. First, the equation is non-linear; however, it is only weakly non-linear for low noise levels. This expression is the gradient of the exponent of the posterior distribution. A globally convergent iterative solution can be assured by gradient descent $\delta S = -\eta\frac{\partial\epsilon}{\partial S}$, with a sufficiently small η. Here, however, we aim for a fast approximate solution for S. Second, the "system matrix", $A^\top A$, may be ill-conditioned or even singular. A useful rewriting that takes care of potential ill-conditioning of the system matrix is,

$$\widehat{S} = \left(A^\top A + \sigma^2\right)^{-1}\left(A^\top X + \sigma^2\left(\widehat{S} - \tanh(\widehat{S})\right)\right) \tag{9.7}$$

which leads to the correct solution in the linear limit if A is near singular. This form furthermore suggests an approximate solution for low noise levels

$$\widehat{S} = S^0 + \sigma^2 H^{-1} \left(S^0 - \tanh S^0 \right)$$
$$S^0 = H^{-1} A^\top X$$
$$H = A^\top A + \sigma^2 \tag{9.8}$$

exposing the fact that the presence of additive noise turns the otherwise linear separation problem into a non-linear one. A non-linear source estimate is also found in Lewicki and Sejnowski's analysis of the overcomplete problem [14].

In the numerical experiments reported below we illustrate the difference between the estimate obtained through equation 13.12 and that obtained by the classical ICA solution (equivalent to S^0). We show that the non-linear solution indeed provides an improved estimate of the sources.

9.4   Joint Estimation of Sources, Mixing Matrix, and Noise Level

Since the likelihood is of the hidden-Gibbs form we can use a generalized Boltzmann learning rule (see [10] and the appendix) to find the gradients of the likelihood of the parameters A, σ^2,

$$\frac{\partial \log L}{\partial A} = \langle \frac{\partial \epsilon}{\partial A} \rangle_{\text{free}} - \langle \frac{\partial \epsilon}{\partial A} \rangle_{\text{clamp}} \tag{9.9}$$

$$\frac{\partial \log L}{\partial \sigma^2} = \langle \frac{\partial \epsilon}{\partial \sigma^2} \rangle_{\text{free}} - \langle \frac{\partial \epsilon}{\partial \sigma^2} \rangle_{\text{clamp}} \tag{9.10}$$

where ϵ is the "energy function" for the Gibbs distribution,

$$\epsilon = \frac{1}{2\sigma^2} \text{Tr}_t((X - AS)^\top (X - AS)) + \sum_{t,m} \log \cosh S \tag{9.11}$$

The averages are calculated w.r.t. the joint distribution $P(X, S)$ for the "free" average and w.r.t. the mixed joint distribution $P_0(X)P(S|X)$ for the "clamped" case, respectively. $P_0(X)$ is the empirical distribution associated with the training set.

The operators to be averaged are given by,

$$\frac{\partial \epsilon}{\partial A} = \frac{1}{\sigma^2} \left(ASS^\top - XS^\top \right) \tag{9.12}$$

$$\frac{\partial \epsilon}{\partial \sigma^2} = -\frac{1}{2\sigma^4} \text{Tr}_t (X - AS)^\top (X - AS) \tag{9.13}$$

The free average is the derivative of the normalization constant of the joint distribution $P(X, S)$, which in the present model only depends on the noise parameter (σ^2). Hence, the free average of the operator conjugate to A vanishes and we need only to calculate expectations, $\langle S \rangle, \langle SS^{\mathsf{T}} \rangle$, for *fixed* observed signals (clamped average). In the general case these expectations are only accessible via simulations. Peterson and Anderson [13] suggested the "Mean Field" approximation (MFA) for the Boltzmann learning problem,

$$\langle S \rangle \approx \widehat{S} \tag{9.14}$$

$$\langle SS^{\mathsf{T}} \rangle \approx \widehat{S}\widehat{S}^{\mathsf{T}} \tag{9.15}$$

Using the MFA we find the estimates $\widehat{A}, \widehat{\sigma^2}$ to be determined by,

$$0 = \frac{1}{\widehat{\sigma^2}} \left(\widehat{A}\widehat{S}\widehat{S}^{\mathsf{T}} - X\widehat{S}^{\mathsf{T}} \right) \tag{9.16}$$

$$0 = \frac{DN}{\widehat{\sigma^2}} - \frac{1}{2\widehat{\sigma^2}^2} \mathrm{Tr}_t (X - \widehat{A}\widehat{S})^{\mathsf{T}} (X - \widehat{A}\widehat{S}) \tag{9.17}$$

The approximation in equation 9.17 neglects the covariance contribution which may be crucial if the estimated source matrix becomes ill-conditioned. Hence, we augment the estimate by a regularization constant $\langle SS^{\mathsf{T}} \rangle \rightarrow \widehat{S}\widehat{S}^{\mathsf{T}} + \beta\mathbf{1}$. β represents a "lumped" effect of the covariance contribution. $\mathbf{1}$ is an $M \times M$ unit matrix. The recursive estimates of A and σ^2 then become,

$$\widehat{A} = X\widehat{S}^{\mathsf{T}} \left(\widehat{S}\widehat{S}^{\mathsf{T}} + \beta\mathbf{1} \right)^{-1} \tag{9.18}$$

$$\widehat{\sigma^2} = \frac{1}{DN} \mathrm{Tr}_t (X - \widehat{A}\widehat{S})^{\mathsf{T}} (X - \widehat{A}\widehat{S}) \tag{9.19}$$

Fluctuation corrections (hence the magnitude of β) can be derived in the low noise limit, based on a Gaussian approximation to the likelihood. The fluctuation expansion to second order (for the neighbourhood of the mode $\widehat{S}$) of the energy functional reads $\epsilon \approx \epsilon_0 + \epsilon_1 + \epsilon_2$, with

$$\epsilon_0 = \frac{1}{2\sigma^2} \mathrm{Tr}_t ((X - A\widehat{S})^{\mathsf{T}} (X - A\widehat{S})) \sum_{t,m} \log \cosh \widehat{S} \tag{9.20}$$

$$\epsilon_1 = -\frac{1}{\sigma^2} \mathrm{Tr}_m (A^{\mathsf{T}} (X - A\widehat{S}) \Delta S^{\mathsf{T}}) + \sum_{t,m} (\tanh \widehat{S}) . * \Delta S \tag{9.21}$$

$$\epsilon_2 = \frac{1}{2\sigma^2} \mathrm{Tr}_t \Delta S^{\mathsf{T}} A^{\mathsf{T}} A \Delta S + \frac{1}{2} \sum_{t,m} (1 - \tanh^2 \widehat{S}) . * \Delta S . * \Delta S \tag{9.22}$$

where .* is used to indicate element-wise multiplication of matrices and where $\widehat{S}$ is the estimate found above. The first order term ϵ_1 is zero by definition of $\widehat{S}$. An approximate expression for β can be obtained from the second order term,

$$\beta^{-1} \approx \frac{1}{M\sigma^2}\mathrm{Tr}A^{\top}A + \frac{1}{MN}\sum_{t,m}(1-\tanh^2\widehat{S}) \tag{9.23}$$

9.5 Simulation Example

We illustrate the performance of the proposed ICA scheme on simulated data. The first experiment is designed to study the role of the non-linear source estimate. We study a data matrix X of dimension $D \times N$, with $D = 4$ sensors and $N = 200$ observations. The signal is created as in equation 13.5 with variable noise levels and $M = 3$ sources. The mixing matrix A was chosen randomly with i.i.d. unit variance and normal components. The "true" sources S_{true} were created from white zero mean, unit variance Gaussian sequences. The individual signal values of two first source sequences were raised to power 2.0 and the sign multiplied back on, while the absolute values of the third source sequence were raised to the power 0.5 and then multiplied by the sign of the original Gaussian variable.

The histograms of the source signals with different *kurtosis* are provided in figure 9.2.

Next, four estimates are evaluated using the given instances of A, σ^2. In particular, we evaluate the low-noise estimate given in equation 13.12, the "linear" estimate S^0 and finally two estimates obtained by performing $Q = 5$ or $Q = 30$ iterations of the non-linear equation equation 13.11. In figure 9.1 we quantify the ability of the solution to estimate the source signals, given that we provide the correct value of A and σ^2. We find that the single iteration explicit solution provides a significantly better estimate of the sources than using the linear solution S^0. In figure 9.3 a qualitative impression is provided by scatterplots of the various estimators versus the true source signals for an experiment at the highest noise level of figure 9.1. In figure 9.2 we show the histograms of the true source signals revealing the super-Gaussian density of the two first signals, and the sub-Gaussian nature of the third signal.

To illustrate the viability of the A, σ^2 learning algorithm we initialize a mixing matrix as a $D \times M$ Gaussian zero mean, unit variance, component random matrix. The noise variance parameter is initialized as $1/100$ of the mean signal variance. Ten thousand iterations of Boltzmann learning is then applied to the two parameters, A, σ^2, as illustrated in figures 9.4 - 9.5. The estimates converged and inspection of the estimated sources shows that very close approximations were obtained. The scatterplots in Figure 9.6 between the three estimated source signals and the three true source signals show that

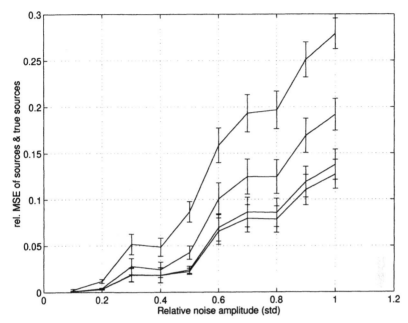

Fig. 9.1. The normalized "Mean Square Error" of the estimator w.r.t. to the known, true, source signals as function of noise variance. The error is normalized by the variance of the true source signals. In terms of performance, i.e., from below: S^{30}, S^5, S^1, S^0. We report the average and std of the average for 50 repetitions.

even for the difficult sub-Gaussian component quite satisfactory separation has been obtained.

For reference we show a similar scatterplot in figure 9.7 , but now for the three principal components that carry the highest variances in a principal component analysis of the same data set. Although these signals are by definition uncorrelated on the training sample, they provide a rather poor approximation to the separation problem.

9.6   Generalization and the Bias-Variance Dilemma

The parameters of our blind separation model are estimated from a finite sample and therefore they are stochastic functions of the training set noise. This means that the generalization ability is of key interest: can we expect the model and the parameters to perform well on independent test data? Within the likelihood formulation the generalization error of a specific set of parameters and model is given by the average negative log-likelihood,

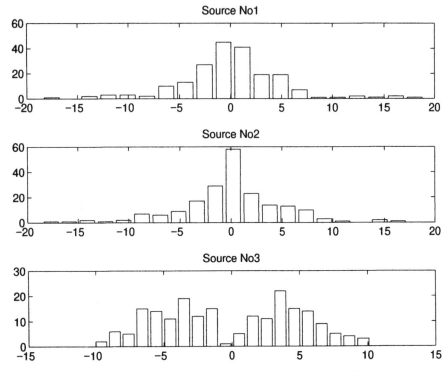

Fig. 9.2. Histograms of the $M = 3$ true source signals used in the numerical experiment. The two upper sources have positive kurtosis distributions, while the last source is drawn from a negative kurtosis distribution.

$$\Gamma(A, \theta, \psi) = \int P_*(X)[-\log P(X|A, \theta, \psi)]dX$$

$$\int P_*(X)[-\log \int dSP(X - AS|\theta)P(S|\psi)]dX \qquad (9.24)$$

cf. equation (13.6). $P_*(X)$ is the true distribution of the data and, if evaluated on a test set, the empirical distribution is

$$P_{\text{emp}}(x) = \frac{1}{N}\sum_{n=1}^{N}\delta(x - x_n) \qquad (9.25)$$

Generalization errors for unsupervised learning schemes (principal components and clustering) are discussed in [7]. The generalization error is a principled tool for model selection. In the context of blind separation the optimal number of sources in the separation is of crucial interest. We face

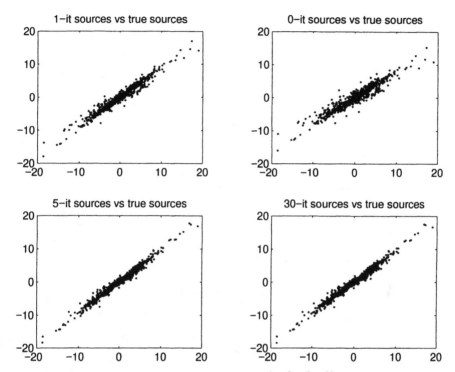

Fig. 9.3. Scatterplots of the four "solutions" (S^1, S^0, S^5, S^{30}, respectively) versus the true sources for the maximal noise level used in Figure 1.

a typical bias - variance dilemma [5,21]. If too few components are used, a structured part of the signal will be lumped with the noise, hence leading to a high generalization error because of "lack of fit". On the other hand, if too many sources are used, we expect "overfit" since the model will use the additional degrees of freedom to fit non-generic details in the training data.

To illustrate generalizability in the context of ICA, we generated a simulated data set along the lines now with $D = 8$ and a "true" number of super-Gaussian sources $M_{\text{true}} = 3$. Models with $M = 2, ..., 7$ sources were adapted and the generalization error was estimated on a test set of another $N_{\text{test}} = 1000$ point generated using the same procedure as we used for the training set. The variation of the generalization error as model dimension was increased is shown in figure 9.8. As expected we found a bias - variance trade-off with excess generalization error for large and small models, while the optimal model size was indeed $M = M_{\text{true}}$. Among ten repetitions of the experiment, nine picked $M = M_{\text{true}} = 3$, while one experiment had minimal generalization error at $M = 4$. In general we expect a distribution of "optimal models" which is peaked at the "true" model dimension and with a width that decreases with increasing sample and test set sizes.

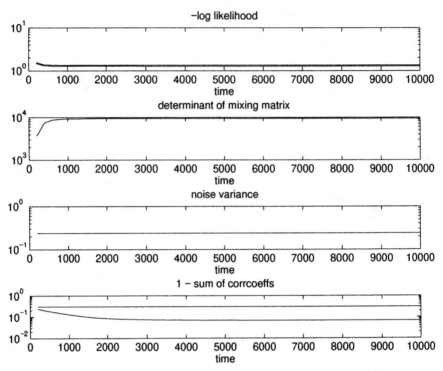

Fig. 9.4. Convergence of Boltzmann Learning for a mixing problem based on the sources of figure 9.3. The first panel shows the evolution of the negative log-likelihood on the training sequence and on a test sequence. The second panel shows the evolution of the determinant of the system matrix $A^T A$. The third panel shows the rapid convergence of the noise variance estimate, while the fourth panel shows the temporal development of the average correlation coefficients between the true sources and the estimated sources (we show $1 - \frac{1}{3} \sum \text{corrcoef}$). For reference we show the average correlation coefficients for the sources estimated by principal component analysis of the training sequence (horizontal line).

9.7   Application to Neuroimaging

Principal component analysis is a popular tool for analysis of short image sequences. The use of PCA in functional neuroimaging has been pioneered by Moeller and Strother in the so-called SSM model [17]. Independent component analysis has been pursued in this context by McKeown et al. [16]. In [16] the fMRI signals were analyzed for independent *spatial patterns*. Here we use our ICA scheme in the form discussed above, looking for independent temporal patterns, and we compare the independent components and their spatial patterns with those obtained by PCA. The basic tool for PCA is singular value decomposition (SVD). In the SVD of a short image sequence $(N < M)$, X is decomposed,

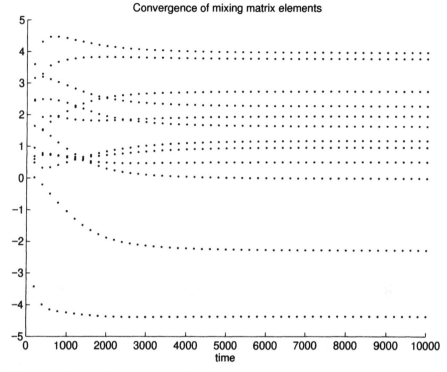

Fig. 9.5. Temporal evolution of the $D \times M = 12$ matrix elements of the estimated mixing matrix.

$$X = U\Lambda V^\top \tag{9.26}$$

where U is a $M \times N$ matrix, while Λ, V are $N \times N$ matrices. Λ is a diagonal matrix with non-negative ordered elements (the largest first)– the singular values. U are eigenvectors corresponding to the N non-zero eigenvalues of $XX^\top$, while V is an orthogonal matrix of the eigenvectors of $X^\top X$,

$$XX^\top = U\Lambda V^\top V\Lambda U^\top = U\Lambda^2 U^\top \tag{9.27}$$
$$X^\top X = V\Lambda U^\top U\Lambda V^\top = V\Lambda^2 V^\top \tag{9.28}$$

hence Λ holds the square roots of the corresponding eigenvalues. Viewed as temporal sequences, $\Lambda V^\top$ are "uncorrelated source signals". The associated spatial patterns are orthogonal, being eigenvectors of a symmetric real matrix. However, as noted above we are often interested in a slightly more general separation of image sources that are independent in time, but not

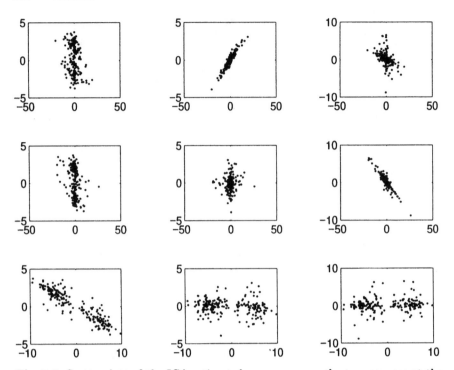

Fig. 9.6. Scatterplots of the ICA estimated sources versus the true sources at the end of the estimation process.

necessarily orthogonal in space, i.e., we would like to be able to perform a more general decomposition of the data matrix, corresponding to the model in equation 13.5. For short image sequences we can make use of the SVD for simplification of the ICA problem. The approach taken here is similar to the so-called "cure for extremely ill-posed learning" [12] used to simplify supervised learning in short image sequences.

We first note that the likelihood, considered a function of the columns of A (spatial patterns), can be split into two parts: part A_1 is orthogonal to the subspace spanned by the N rows of X and part A_2 is situated in the subspace spanned by the N columns of X. The first is part trivially minimized for any non-zero configuration of sources by putting $A_1 = 0$. It simply does not "couple" to data. The remaining part A_2 can be written as UB in terms of an $N \times N$ matrix B, simplifying the mixing model for the short sequence problem to,

$$Y \equiv U^\top X = BS + \Gamma \tag{9.29}$$

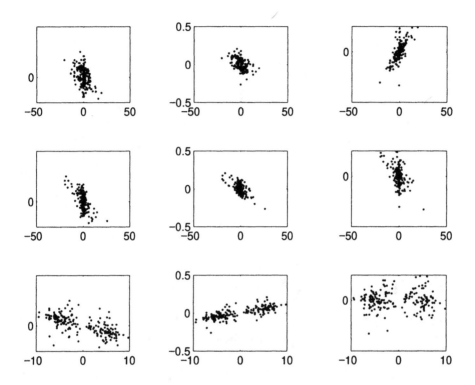

Fig. 9.7. For reference we show the scatterplots of the PCA estimated sources versus the true sources.

with Y, Γ being $N \times N$ matrices. We note that it often may be possible to further limit the dimensionality of the SVD subspace retained in equation 9.29, hence further reducing the "sensor" dimensionality, D, of the remaining problem.

We use the above scheme for the analysis of a functional Magnetic Resonance Imaging sequence, acquired by Dr. Egill Rostrup, Hvidovre Hospital, Denmark, as part of a study involving visual stimulation. A single slice, holding 91×71 pixels, through the visual cortex was acquired with a short time interval between successive scans of TR $= 333$ msec. Visual stimulation in the form of a flashing annular checkerboard pattern was interleaved with periods of fixation. A "run" consisting of 25 scans of rest, 50 scans of stimulation, and 25 scans of rest was repeated 10 times. For this analysis a contiguous mask was created with 2440 pixels holding the essential parts of the slice including the visual cortex. SVD was performed on a subset of three runs ($N = 300$) providing matrices $U \sim 2440 \times 300$, $\Lambda \sim 300 \times 300$ and $V \sim 300 \times 300$.

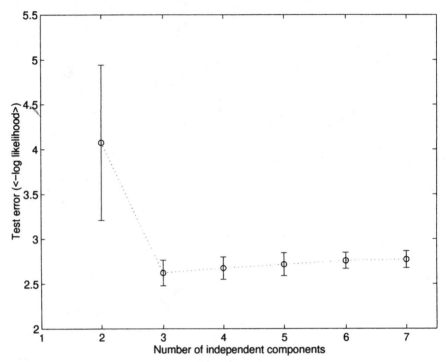

Fig. 9.8. Bias-variance tradeoff for blind separation as model dimensionality is varied. The simulated data set was generated with $M_{\text{true}} = 3$ sources. Smaller or larger models have excess generalization error because of "lack-of-fit" and "overfit" respectively.

In figures 9.9 and 9.10 we show the time course of the nine most significant principal components and their associated eigenimages. A distinct activation response is picked up in the first component and the thresholded eigenimage associated with it is dominated by a cluster of pixels in the visual cortex area as expected. Principal components 3 and 4 show evidence of "non-stationarity" pointing to the third run ($t = 201 - 300$) being different from the two first runs, however the spatial pattern does not provide much clue to the nature of these differences. In figures 9.11 and 9.12 we show the time courses and spatial patterns of the ICA solution. The differentiation of the first two runs from the third run is very prominent in the estimated source signals. Components 3, 7 and 8 are activated in the two first runs, while components 1, 5 and 9 are activated in the third run. Comparing the spatial patterns of component 1 (active in the two first runs) and component 8 (active in the third run) we find that there are small areas anterior to the main activation hotspot in the visual cortex which are correlated with the activity of the visual cortex (both are found in the positive excursion set) for

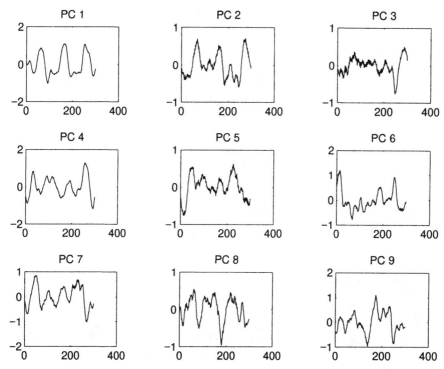

Fig. 9.9. Time courses of the principal components corresponding to the nine largest covariance eigenvalues for a sequence of 300 fMRI scans of a visually stimulated subject. Stimulation takes places at scan times $\tau = 25 - 75$, $\tau = 125 - 175$, and $\tau = 225 - 275$ with a time interval between scans of $TR = 333msec$. The time courses have been smoothened for presentation, eliminating noise and high-frequency physiological components. Note that most of the response is captured by the first principal components, showing a strong response to all three periods of stimulation.

the first two runs, and anti-correlated in the third run (the main area is in the positive excursion set, while the smaller regions are now coloured blue and hence belong to the negative excursion set). Such subtle differentiation of the temporal structure of the fMRI signal is not seen in PCA, presumably due the additional spatial orthogonality constraint.

9.8   Conclusion

A likelihood-MAP formulation of the noisy separation problem has been given. In the face of additive noise, the MAP estimate of the independent sources is a non-linear functional of the observed signal, in contrast with the linear "classical" solution. A numerical experiment showed that the non-linear

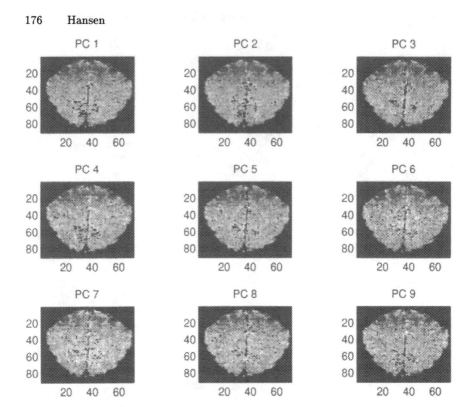

Fig. 9.10. Covariance eigenimages corresponding to the nine most significant principal components. The eigenimage corresponding to the dominating first PC is, as expected, focused in the region of the primary visual cortex. The image is formed by thresholding the eigenimages, so that the 3% fractiles (positive: red, negative: blue) are shown in the context of the averaged fMRI image.

MAP estimate indeed provides a closer approximation to the true source signals in the presence of additive white noise.

As the signal model distribution takes the form of a hidden variable Gibbs distribution, adaptation of the mixing matrix and the noise parameters can be implemented using the Boltzmann learning rule. Experiments on simulated data showed that the learning rule was capable of finding both the mixing matrix and an estimate of the noise level.

In real-world applications of adaptive systems *generalizability* is of key importance. Generalizability depends on the sample size and on model complexity. A relevant complexity parameter for ICA based on the simple prior distribution in equation 13.8 is the number of sources. We proposed the generalization error – defined as the average negative log-likelihood and estimated on an independent test set – as a means for optimization of source model

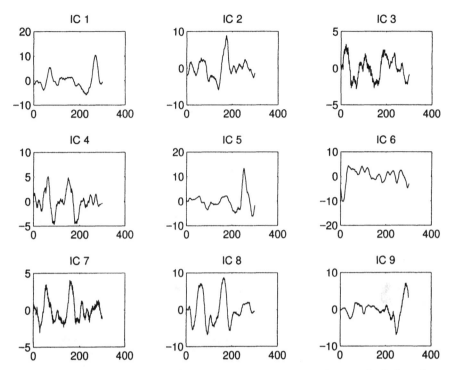

Fig. 9.11. Time courses of the independent components of an analysis based on nine sources. Stimulation takes places at scan times $\tau = 25 - 75$, $\tau = 125 - 175$, and $\tau = 225 - 275$ with a time interval between scans of $TR = 333msec$. The time courses have been smoothed for presentation, eliminating noise and high-frequency physiological components. In contrast to the PCA results in figure 12, we find here that the response is distributed across several sources. IC numbers 3, 7 and 8 have picked up most of the response for the first two runs, while ICs 1, 5 and 9 are dominated by response in the last of the three runs. This more detailed decomposition is possible because the spatial patterns are not subject to the orthogonality constraint. Hence, while the spatial patterns corresponding to the third set of runs are quite similar to those associated with the first two, the ICA solution has differentiated them spatially.

complexity. An experiment on simulated data showed that the estimated generalization error was able to identify the correct source dimensionality.

In an application to short image sequences we simplified the ill-posed modeling problem considerably by use of the singular values decomposition. Furthermore, it was shown that the ICA was able to differentiate three repetitions of the experiment, using its more flexible definition of spatial patterns. The corresponding PCA was not able to capture this apparent non-stationary aspect of the experiment.

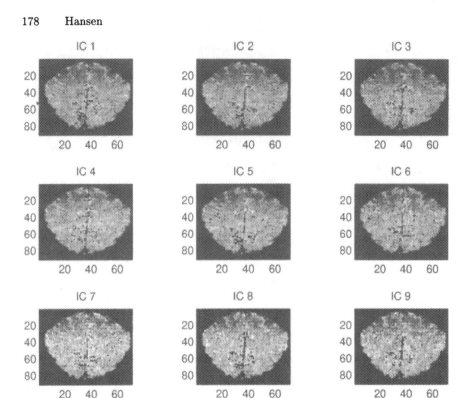

Fig. 9.12. Spatial patterns corresponding to the independent components of figure 9.11. We note the similarities of the patterns corresponding to components 3, 7 and 8 (holding the response to the first two runs) and components 1, 5 and 9 (holding the response for the third run). The ICA algorithm can separate temporal structures that are associated with similar, hence overlapping, but different spatial systems. The image is formed by thresholding the eigenimages, so that the 3% fractiles (positive: red, negative: blue) are shown in the context of the averaged fMRI image.

9.9   Acknowledgments

I thank Thomas Kolenda, Jan Larsen, Te-Won Lee, Michael Lewicki, Martin McKeown, Niels Mørch and Terry Sejnowski for valuable comments and discussions. The author was supported in part by a McDonnell-Pew Center travel grant. This research was carried out while the author was visiting the Brain Image Analysis Lab (BIAL). I wish to thank Terry Jernigan and the BIAL staff and researchers for creating an enjoyable atmosphere for the visit.

9.10   Appendix: The Generalized Boltzmann Learning Rule

The Boltzmann Learning rule was the first learning rule for neural networks with hidden units and was formulated for binary stochastic units [10]. The main virtue from a modern practical point of view is that it allows learning in Gibbs distributions with hidden variables *without computing the normalization integral.* Here I provide the general gradient calculation. A systematic second order (Newton) scheme is also available but is not necessary for this context, since we can solve the stationarity equations directly. The generalized Boltzmann rule was earlier used in [8] for parameter estimation in Random Markov Fields.

Let $T = \{x_n | n = 1, ..., N\}$ be a sample of a stochastic variable with distribution $P_*(x)$. Let our model distribution be of the hidden variable Gibbs form,

$$P(x|\theta) = \int dS P(x, S|\theta) \tag{9.30}$$

with

$$P(x, S|\theta) = Z^{-1} \exp\left(-\epsilon(x, S|\theta)\right) \tag{9.31}$$

$$Z = \int dx dS \exp\left(-\epsilon(x, S|\theta)\right) \tag{9.32}$$

The objective function for the parameters θ is the negative log-likelihood function,

$$C = \int dx P_0(x)[-\log P(x|\theta)] \tag{9.33}$$

where the distribution $P_0(x)$ can be either $P_*(x)$, i.e., C is the generalization error, or

$$P_0(x) = P_{\text{emp}}(x) = \frac{1}{N} \sum_{n=1}^{N} \delta(x - x_n) \tag{9.34}$$

in which case C is the training error. The objective function can be augmented with a log-prior term leading to a penalized maximum likelihood estimation.

We estimate parameters by minimization of C, hence we look for the gradients $\frac{\partial C}{\partial \theta}$. Computing the derivative we face two separate terms,

$$\frac{\partial C}{\partial \theta} = -\frac{\partial \log Z}{\partial \theta} + \frac{\partial}{\partial \theta} \int dS \exp\left(-\epsilon(x, S|\theta)\right) \tag{9.35}$$

The first derivative on the left hand side of equation 9.35 is

$$-\frac{\partial \log Z}{\partial \theta} = -\frac{1}{Z} \int dx dS \frac{\partial}{\partial \theta} e^{-\epsilon} \tag{9.36}$$

$$= \int dx dS \frac{\partial \epsilon}{\partial \theta} P(x, S|\theta) \tag{9.37}$$

$$\equiv \langle \frac{\partial \epsilon}{\partial \theta} \rangle_{\text{free}} \tag{9.38}$$

The second term on the left hand side of equation 9.35 is

$$\int dx P_0(x) \frac{\partial}{\partial \theta} \log \int dS e^{-\epsilon} = \int dx P_0(x) \frac{1}{Z P(x)} \int dS \frac{\partial}{\partial \theta} e^{-\epsilon} \tag{9.39}$$

$$= -\int dx P_0(x) \int dS \frac{\partial \epsilon}{\partial \theta} \frac{P(x, S|\theta)}{P(x)} \tag{9.40}$$

$$= -\int dx P_0(x) \int dS \frac{\partial \epsilon}{\partial \theta} P(x|S, \theta) \tag{9.41}$$

$$\equiv -\langle \frac{\partial \epsilon}{\partial \theta} \rangle_{\text{clamped}} \tag{9.42}$$

References

1. S. Amari, A. Cichocki, and H. Yang. A new learning algorithm for blind signal separation. In D. Touretzky, M. Mozer, and M. Hasselmo, editors, *Advances in Neural Information Processing Systems*, **8**, 757–763, MIT Press, Cambridge MA, 1996.
2. H. Attias and C.E. Schreiner. Blind source separation and deconvolution by dynamic component analysis. *Neural Networks for Signal Processing VII: Proceedings of the 1997 IEEE Workshop*, 1997.
3. A. J. Bell and T. J. Sejnowski. An information-maximization approach to blind separation and blind deconvolution. *Neural Computation*, **7**(6):1129–1159, 1995.
4. A. Belouchrani and J.-F. Cardoso. Maximum likelihood source separation by the expectation-maximization technique: deterministic and stochastic implementation. *In Proc. NOLTA*, 49-53, 1995.
5. S. Geman, E. Bienenstock, and R. Doursat. Neural Networks and the Bias/Variance Dilemma, *Neural Computation*, **4**: 1-58, 1992.
6. P. Comon. Independent component analysis – a new concept? *Signal Processing*, **36**:287–314, 1994.
7. L.K. Hansen and J. Larsen. Unsupervised learning and generalization. *Proceedings of the IEEE International Conference on Neural Networks*.1: 25 - 30, 1996.
8. L.K. Hansen, L. Nonboe Andersen, U. Kjems, J. Larsen. Revisiting Boltzmann learning: parameter estimation in Markov random fields. *Proceedings of International Conference on Acoustics Speech and Signal Processing*. **6**: 3395 - 3398, 1996.

9. L.K. Hansen, J. Larsen, F.Å. Nielsen, S.C. Strother, E. Rostrup, R. Savoy, N. Lange, J.J. Sidtis, C. Svarer, O.B. Paulson. Generalizable patterns in neuroimaging: how many principal components? *NeuroImage*. **9**: 534 - 544, 1999.

10. G. E. Hinton and T. J. Sejnowski. *Learning and relearning in Boltzmann machines* In D.E. Rumelhart and J.L. McClelland, Eds. *Parallel Distributed Processing: Explorations in the Microstructure of Cognition*. **1**: 282, MIT Press, Cambridge, 1986.

11. C. Jutten and J. Herault. Blind separation of sources: An adaptive algorithm based on neuromimetic architecture. *Signal Processing*. **24**: 1 - 10, 1991.

12. B. Lautrup, L.K. Hansen I. Law, N. Mørch, C. Svarer, S.C. Strother. Massive weight sharing: A cure for extremely ill-posed problems. H.J. Hermanet al., eds. *Supercomputing in Brain Research: From Tomography to Neural Networks*. 137 - 148, 1995.

13. T.-W. Lee. Independent component analysis: theory and applications, *Kluwer Academic Publishers*, 1998.

14. M. S. Lewicki and T. J. Sejnowski. Learning overcomplete representations *Neural Computation*. **12**: 2, 2000.

15. D. MacKay. Maximum likelihood and covariant algorithms for independent components analysis. "Draft 3.7", 1996.

16. M. J. McKeown, T. P. Jung, S. Makeig, G. Brown, S. S. Kindermann, T.-W. Lee and T. J. Sejnowski. Spatially independent activity patterns in functional magnetic resonance imaging data during the stroop color-naming task. *Proceedings of the National Academy of Sciences USA*, **95**: 803 - 810, 1998.

17. J. R. Moeller and S. C. Strother. *A regional covariance approach to the analysis of functional patterns in positron emission tomographic data*. J. Cereb. Blood Flow Metab. **11**: A121-A135 (1991).

18. L. Molgedey & H. Schuster. Separation of independent signals using time-delayed correlations. *Physical Review Letters*. **72**: 3634 - 3637, 1994.

19. E. Moulines, J.-F. Cardoso, E. Gassiat. Maximum likelihood for blind separation and deconvolution of noisy signals using mixture models. *Proc. ICASSP, Munich*, **5**: 3617 - 3620, 1997.

20. N. Mørch, U. Kjems, L.K. Hansen, C. Svarer, I. Law, B. Lautrup, S.C. Strother, and K. Rehm. Visualization of neural networks using saliency maps. *Proceedings of 1995 IEEE International Conference on Neural Networks*, 2085 - 2090, 1995.

21. N. Mørch, L.K. Hansen, S.C. Strother, C. Svarer, D.A. Rottenberg, B. Lautrup, R. Savoy, O.B. Paulson. Nonlinear versus linear models in functional neuroimaging: learning curves and generalization crossover. *Proceedings of the 15th International Conference on Information Processing in Medical Imaging*. **1230**: 259 - 270, 1997.

22. E. Oja. PCA, ICA, and nonlinear Hebbian learning. *Proc.Int.Conf. on Artificial Neural Networks*, 89 - 94, 1995.

23. B.A. Olshausen. Learning linear, sparse, factorial codes. *A.I. Memo 1580, Massachusetts Institute of Technology*, 1996.

24. B. A. Pearlmutter and L. C. Parra. A context-sensitive generalization of ICA. *Proc. International Conference on Neural Information Processing*, 1996.

25. C. Peterson & J.R. Anderson. A mean field theory learning algorithm for neural networks. *Complex Systems*. **1**: 995 - 1019, 1987.

10 Searching for Independence in Electromagnetic Brain Waves

Ricardo Vigário, Jaakko Särelä and Erkki Oja

10.1 Introduction

Multichannel recordings of the electromagnetic fields emerging from neural currents in the brain generate large amounts of data. Suitable feature extraction methods are, therefore, useful to facilitate the representation and interpretation of the data.

Independent Component Analysis (ICA) has been shown to be an efficient tool for artifact identification and extraction from electroencephalographic (EEG) and magnetoencephalographic (MEG) recordings. In addition, ICA has been applied to the analysis of brain signals evoked by sensory stimuli. This chapter reviews our recent results in this field. The exposé is mainly based on [38,40].

With the advent of new anatomical and functional imaging methods, it is now possible to collect vast amounts of data from the living human brain. It has thus become very important to extract the essential features from the data to allow an easier representation or interpretation of their properties. Traditional approaches to solving this feature extraction or dimension reduction problem include, e.g., principal component analysis (PCA), projection pursuit, and factor analysis. This chapter focuses on a novel signal processing technique, independent component analysis (ICA), which allows blind separation of sources, linearly mixed at the sensors, assuming only the statistical independence of these sources.

Electroencephalograms (EEG) and magnetoencephalograms (MEG) are recordings of electric and magnetic fields of signals emerging from neural currents within the brain. The challenges presented to the signal processing community by the researchers employing EEG and MEG include the identification and removal of artifacts from the recordings and the analysis of the brain signals themselves.

In section 10.2 we present a short description of the independent component analysis theory, together with an algorithm capable of performing such analysis. Section 10.3 reviews the origins of EEG and MEG techniques. In Section 10.4, we validate the application of the ICA model to EEG and MEG. The use of ICA for identification of artifacts in EEG and MEG as well as the decomposition of event-related activity is presented in section 10.5.

10.2   Independent Component Analysis

10.2.1   The Model

ICA [2,8,10,15,23] is a novel statistical technique that aims at finding linear projections of the data that maximize their mutual independence. Its main applications are in feature extraction [14,31], and blind source separation (BSS) (for excellent reviews, see [2,8], with special emphasis to physiological data analysis [37,39,42,22,3,28], and audio signal processing [36]).

As in many other linear transformations, it is assumed that at time instant k the observed n-dimensional data vector, $\mathbf{x}(k) = [x_1(k), \ldots, x_n(k)]^T$ is given by the models:

$$x_i(k) = a_{i1}s_1(k) + a_{i2}s_2(k) + \cdots + a_{im}s_m(k)$$

or, in a more compact notation,

$$\mathbf{x}(k) = \sum_{j=1}^{m} \mathbf{a}_j s_j(k) = \mathbf{A}\mathbf{s}(k). \tag{10.1}$$

The source signals, $s_1(k), \ldots, s_m(k)$, are supposed to be stationary independent but unknown, as are the coefficients of the mixing matrix $\mathbf{A} = [\mathbf{a}_1, \ldots, \mathbf{a}_m]$. Furthermore, only up to one source may be Gaussian. The goal is to estimate both unknowns from a sample of the $\mathbf{x}(k)$, with appropriate assumptions on the statistical properties of the source distributions. The solution is sought in the form

$$\hat{\mathbf{s}}(k) = \mathbf{B}\mathbf{x}(k) \tag{10.2}$$

where $\mathbf{B}$ is called the separating matrix.

The general BSS problem requires $\mathbf{A}$ to be an $n \times m$ matrix of full rank, with $n \geq m$ (i.e., there are at least as many mixtures as the number of independent sources). In most algorithmic derivations, an equal number of sources and sensors is assumed.

In equation 10.1, schematically illustrated in figure 10.1, we omit additive noise; some analysis of the noisy model can be found in [18,38].

10.2.2   The FastICA Algorithm

In the FastICA algorithm, the initial step is whitening or sphering. By a linear transformation, the measurements $x_i(k)$ and $x_j(k)$, for all i, j, are made uncorrelated and unit-variance [16]. The whitening facilitates the separation of the underlying independent signals [24]. In [17], it has been shown that a well-chosen compression, during this stage, may be necessary in order to reduce the overlearning (overfitting), typical of ICA methods. The result of a

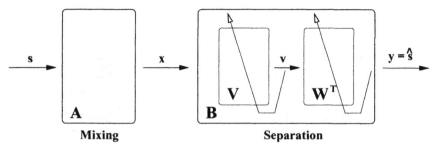

Fig. 10.1. Schematical illustration of the mathematical model used to perform the ICA decomposition.

poor compression choice is the production of solutions practically zero almost everywhere, except at the point of a single spike or bump.

The whitening may be accomplished by PCA projection: $\mathbf{v}(k) = \mathbf{V}\mathbf{x}(k)$, with $E\{\mathbf{v}(k)\mathbf{v}(k)^T\} = I$. The whitening matrix $\mathbf{V}$ is given by $\mathbf{V} = \mathbf{\Lambda}^{-1/2}\mathbf{U}^T$, where $\mathbf{\Lambda} = \text{diag}[\lambda(1),\ldots,\lambda(m)]$ is a diagonal matrix with the eigenvalues of the data covariance matrix $E\{\mathbf{x}(k)\mathbf{x}(k)^T\}$, and $\mathbf{U}$ is a matrix with the corresponding eigenvectors as its columns. The transformed vectors $\mathbf{v}(k)$ are called white or sphered, because all directions have the same unit variance.

In terms of $\mathbf{v}(k)$, the equation 10.1 becomes

$$\mathbf{v}(k) = \mathbf{V}\mathbf{A}\mathbf{s}(k) \tag{10.3}$$

and we can show that matrix $\mathbf{W} = \mathbf{V}\mathbf{A}$ is orthogonal [16]. Therefore, the solution is now sought in the form:

$$\hat{\mathbf{s}}(k) = \mathbf{W}^T\mathbf{v}(k) \tag{10.4}$$

Uncorrelation and independence are equivalent concepts in the case of Gaussian distributed signals. PCA is therefore sufficient for finding independent components in this case. However, PCA is not suited for dealing with non-Gaussian data, where independence is a more restrictive requirement than uncorrelation. Several authors have shown [23,10,4,9,15] that higher-order statistics are required to deal with the independence criterion.

According to the Central Limit Theorem, the sum of m independent random variables, with identical distribution functions approaches the normal distribution as m tends to infinity [32]. Reversing this principle, we may thus replace the problem of finding the independent source signals by a suitable search for linear combinations of the mixtures that maximize a certain measure of non-Gaussianity; for a more rigorous derivation, see [15].

In many ICA algorithms, the fourth-order cumulant, also called the kurtosis, is used as a measure of non-Gaussianity. For the ith source signal, the kurtosis is defined as

$$kurt(s_i) = E\{s_i^4\} - 3[E\{s_i^2\}]^2 \tag{10.5}$$

$E\{\cdot\}$ denotes the mathematical expectation value of the bracketed quantity. The kurtosis is negative for source signals whose amplitude has sub-Gaussian probability densities (distributions flatter than Gaussian), positive for super-Gaussian (sharper than Gaussian, and with longer tails), and zero for Gaussian densities. Maximizing the norm of the kurtosis leads to the identification of non-Gaussian sources.

Consider a linear combination $y = \mathbf{w}^T\mathbf{v}$ of the white random vector $\mathbf{v}$, with $\|\mathbf{w}\| = 1$; see equation 10.4. Then $E\{y^2\} = 1$ and $kurt(y) = E\{y^4\} - 3$, whose gradient with respect to $\mathbf{w}$ is $4E\{\mathbf{v}(\mathbf{w}^T\mathbf{v})^3\}$.

The FastICA [16] is a fixed-point algorithm which finds one of the columns of the separating matrix $\mathbf{W}$ (noted $\mathbf{w}$) and so identifies one independent source at a time. The corresponding independent source signal can then be found using equation (10.4). Each lth iteration of this algorithm is defined as

$$\mathbf{w}^*_l = E\{\mathbf{v}(\mathbf{w}^T_{l-1}\mathbf{v})^3\} - 3\mathbf{w}_{l-1}$$
$$\mathbf{w}_l = \mathbf{w}^*_l/\|\mathbf{w}^*_l\| \tag{10.6}$$

In order to estimate more than one solution, up to a maximum of m, the algorithm may be run repeatedly. It is, nevertheless, necessary to remove the information contained in the solutions already found, to estimate a different independent component each time. For details on the FastICA algorithm, see [16]. Further reading on algorithmic implementations of the ICA technique, as well as its extensions and relations to other data analysis techniques, can be found in [15,38].

All studies reported in this chapter were carried out using MATLAB code, based on the FastICA package [1].

10.3   Electro- and Magnetoencephalography

Several anatomical and functional imaging methods have been developed to study the human brain *in vivo*. Computerized X-ray tomography (CT) and magnetic resonance imaging (MRI), with their very good spatial resolution (around 1 mm), give accurate anatomical images of the brain. Positron emission tomography (PET) and functional MRI (fMRI) provide functional information by probing the changes in metabolic activity in the brain. The time resolution is ultimately limited by the hæmodynamic response in the brain, which is a few seconds. For further reading on these brain mapping techniques, see e.g. [35].

The only non-invasive methods which provide direct information about the neural dynamics on a millisecond scale are electroencephalography (EEG) and magnetoencephalography (MEG) [30,12]. Under favourable conditions the spatial resolution is about 5 mm. These methods are sensitive to the electrical activity within the brain, which constitutes the means by which information is transmitted and processed. The following discussion is based on a comprehensive review [11], with contributions from [33,19].

The extensive use of EEG for monitoring the electrical activity within the human brain, both for research and clinical purposes, has made it one of the most widespread brain mapping techniques to date. It is used both for the measurement of spontaneous rhythmic activity and for the study of evoked potentials (EPs). This triggered activity is time-locked to a particular stimulus that may be e.g. auditory or somatosensory. The left-hand side of

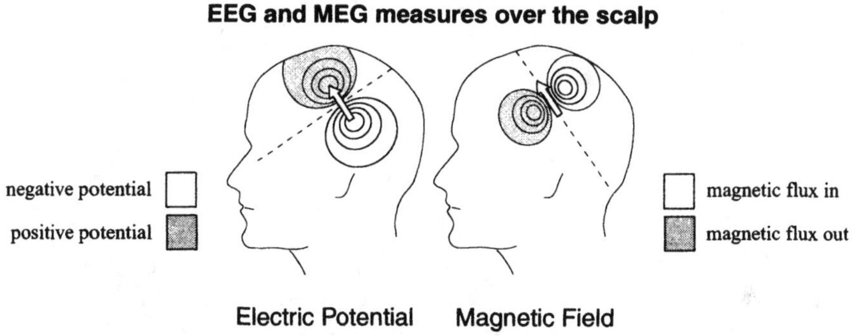

EEG and MEG measures over the scalp

Electric Potential   Magnetic Field

Fig. 10.2. Electric potential and magnetic field associated with or generated by an ideal tangential current dipole, represented by a white arrow, calculated using a spherical head model. Adapted from [11].

figure 10.2 depicts the electric potential distribution caused by a current dipole. The field distributions have been calculated using a spherical model of the head comprising four layers: the brain, the cerebrospinal fluid, the skull, and the scalp [11]. Typical clinical EEG systems use around 20 electrodes, evenly distributed over the head, to measure the potential distribution, which is of the order of one hundred microvolts: evoked potentials may be two orders of magnitude lower than this value. State-of-the-art EEGs may consist of a couple of hundred sensors, yet optimal constructions use between 64 or 128 evenly placed electrodes [26].

The magnetic field, associated with a tangential current dipole, is depicted on the right-hand side of figure 10.2. This field is more local, as it does not suffer from the smearing caused by the different electric conductivities of the several layers between the brain and the measuring devices seen in EEG.

Superconducting QUantum Interference Devices (SQUIDs) are needed to measure the very weak magnetic fields of the brain, typically $50 - 500$ fT; the measurements are carried out inside a magnetically shielded room. A SQUID consists of a superconducting ring interrupted by one or two Josephson junctions [21]. The voltage and resistance over a Josephson junction are zero unless the current through the junction is above the critical current I_c. The

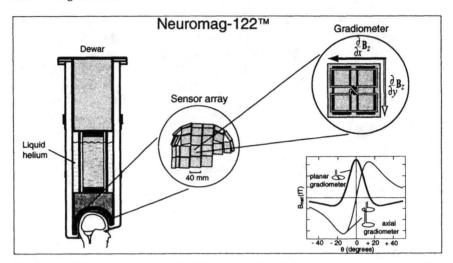

Fig. 10.3. Schematic view of the Neuromag-122 whole scalp magnetometer. The dewar, filled with liquid helium, holds 61 measuring units (middle insert). Each unit contains two figure-of-eight shaped planar gradiometric coils, the outputs of which are proportional to the orthogonal derivatives of the radial magnetic flux (insert on the upper right). The lower right insert shows the amplitudes of the signals detected by axial and planar flux transformers. Adapted from [11].

superconducting characteristics of the device are guaranteed through its immersion in liquid helium, at a temperature of $-269°C$. Figure 10.3 shows the Neuromag-122 (manufactured by Neuromag Ltd., Finland). The whole-scalp sensor array is composed of 122 planar gradiometers, organized in pairs at 61 locations around the head, measuring simultaneously the tangential derivatives $\partial B_z/\partial x$ and $\partial B_z/\partial y$ of the magnetic field component B_z normal to the helmet-shaped bottom of the dewar. The inserts in figure 10.3 show both the layout of the sensor array and one pair of orthogonal planar gradiometers.

The lower right insert in figure 10.3 compares two types of magnetic flux transformers: the planar and the axial gradiometers. Both of them consist of two loops wound in opposite directions to cancel homogeneous background fields. As shown in figure 10.3, the axial gradiometer measures the maximum signal on both sides of the source, whereas the planar gradiometer employed in Neuromag-122 is most sensitive to currents directly beneath it.

10.4   On the Validity of the Linear ICA Model

The application of ICA to the study of EEG and MEG signals assumes that several conditions are verified: the existence of statistically independent source signals, their instantaneous linear mixing at the sensors, and the stationarity of both the source signals and the mixing process.

The independence criterion applies solely to the statistical relations between the amplitude distributions of the signals involved and not to considerations upon the morphology or physiology of certain neural structures. However, the different nature of the sources of the artifacts from those of the actual brain signals has been the driving force behind the application of ICA to the removal of artifacts from EEG and MEG. Analysis of the distributions of artifacts such as the cardiac cycle, ocular activity or a digital watch has shown the statistical independence approximation to be accurate.

Because most of the energy in EEG and MEG signals lies below 1 kHz, the quasistatic approximation of Maxwell equations holds and each time instance can be considered separately [11]. Therefore, there is no need to introduce any time-delays and the instantaneous mixing model is valid.

The non-stationarity of EEG and MEG signals is well documented [6]. When considering the underlying source signals as stochastic processes, the requirement of stationarity is in theory necessary to guarantee the existence of a representative (non-Gaussian) distribution of the sources. Yet, in the implementation of the batch FastICA algorithm, the data are considered as random variables and their distributions estimated from the whole data set. This removes the strict requirement of stationarity.

The stationarity of the mixing process corresponds to the existence of a constant mixing matrix $\mathbf{A}$. This result agrees with widely accepted neuronal source models [34,29].

10.5 The Analysis of EEG and MEG Data

10.5.1 Artifact Identification and Removal from EEG/MEG

A review of artifact identification and removal, with special emphasis on ocular artifacts, can be found in [7,37]. The simplest, and most widely used, method consists of discarding the portions of the recordings containing attributes (e.g. amplitude peak, frequency contents, variance and slope) that exceed a determined threshold. This may lead to significant loss of data, and the complete inability to study interesting brain activity occuring near or during strong eye activity, such as in visual tracking experiments.

Other methods include the subtraction of a regression portion of one or more additional inputs (e.g. from electro-oculograms, electro-cardiograms, or electro-myograms) from the measured signals. This technique is more likely to be used in EEG recordings, but may, in some situations, be applied to MEG. It should be noted that this may lead to the insertion of undesirable new artifacts into the brain recordings [20].

The signal-space projection method has been successfully employed in artifact removal [13]. Another related technique, subtracting the contributions of modeled dipoles accounting for the artifact, has been proposed by [5]. In both methods we need either a good model of the artifact source or a considerable amount of data where the amplitude of the artifact is much higher

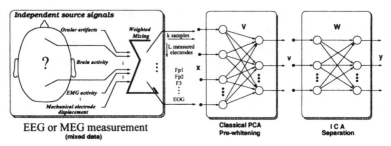

Fig. 10.4. Block diagram of artifact extraction from EEG or MEG recordings, using independent component analysis.

than that of the EEG or MEG. As can be seen in [37,39], those restrictions don't apply to the ICA technique, as several artifacts are extracted in which the signal-to-noise ratio is well below unity (considering as noise everything other than the artifact).

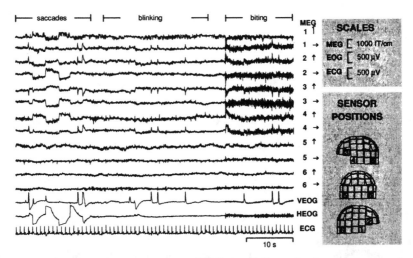

Fig. 10.5. A subset of 12 spontaneous MEG signals from the frontal, temporal and occipital areas. The data contains several types of artifacts, including ocular and muscle activity, the cardiac cycle, and environmental magnetic disturbances.

The assumption of independence of the sources, based on anatomical and physiological considerations, was justified through the successful application of ICA to the identification of artifacts in EEG and MEG, as reported in [37,39] and [22]. The block diagram in figure 10.4 illustrates the use of ICA in the identification and removal of artifacts from EEG and MEG recordings.

Figure 10.5 presents a subset of 12 MEG artifact-contaminated signals, from a total of 122 used in the experiment. Several artifact structures are evident, such as eye and muscle activity. These can be extracted using ICA, as shown in figure 10.6. IC1 and IC2 are clearly activation of two different muscle sets, whereas IC3 and IC5 are, respectively, horizontal eye movements and blinks. Furthermore, other disturbances with weaker signal-to-noise ratio, such as the heart beat and a digital watch, are also extracted (IC4 and IC6, respectively). For each component the left, back and right views of the field patterns are shown. These field patterns are given by the corresponding mixing vector a_i.

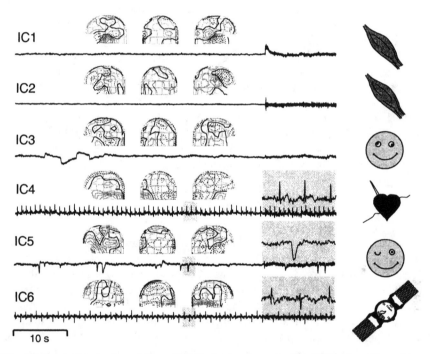

Fig. 10.6. Artifacts found from MEG data using the FastICA algorithm. Three views of the field patterns generated by each independent component are plotted on top of the respective signal. Full lines correspond to magnetic flux exiting the head, whereas the dashed lines correspond to the flux inwards.

A similar approach to the removal of artifacts from cardiographic signals has been presented in [3,43].

10.5.2   Analysis of Multimodal Evoked Fields

State-of-the-art approaches to processing magnetic evoked fields are often based on a careful expert scrutiny of the complete data (in raw format or averaged around the responses to repeating stimuli). At each time instance, one or a set of neural sources, often of a dipolar nature, are modeled in order to produce as good a fit to the data as possible. The quality of the fit is then evaluated through its goodness-of-fit [25]. The choice of the time instances where this fitting should be made, as well as the type of source models employed, are therefore crucial.

The application of ICA in event-related studies was first introduced in the blind separation of auditory evoked potentials by [27]. This method has been further developed in relation to magnetic auditory and somatosensory evoked fields (AFs and SEFs, respectively) in [42,41], using the more efficient FastICA algorithm in a deflationary mode. The following text will review the most important results obtained in our studies. Further information on the experimental setup and particular results can be found in the respective publications.

Without any prior source model assumption, other than the statistical independence from the rest of the MEG signals, the most significant independent components we have found in different modality event-related studies have shown patterns that agree with the conventional dipole approximation. Adding the dipole modeling information to the calculation of the source locations, we have found them to fall on very acceptable brain areas (the difference from conventional methods was well below 1 cm, therefore within the spatial precision of MEG).

In [41], ICA was shown to be able to differentiate between somatosensory and auditory brain responses in the case of vibrotactile stimulation, which, in addition to tactile stimulus, also produced a concomitant sound. In figure 10.7, note that PCA hasn't been able to resolve the complexity of the original MEG signals, most of the components still presenting combined somatosensory and auditory responses (see figure. 10.7b). The field pattern lines, shown in figure 10.7d, correspond to the columns of the estimated mixing matrix associated with the first two independent components in figure 10.7c. Full lines depict the magnetic flux coming out of the head, whereas dashed ones correspond to the flux entering the head. The current dipoles best accounting for these field patterns are also shown. Note that, based on the instantaneous linear ICA model, the field patterns are constant, simplifying the analysis of the results and enabling a semi-automatic processing of the data.

The location of the equivalent current sources fall on the expected brain regions for the particular stimulus. Figure 10.7e shows these brain sources superimposed onto vertical and horizontal MRI slices. The black dots in the MRI correspond to activation of the primary auditory cortex. In addition, the orientations of the dipoles, represented by the lines starting at the dots,

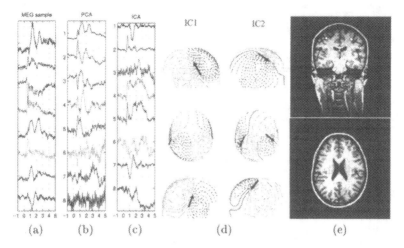

Fig. 10.7. Results of the application of FastICA to averaged brain MEG responses to a vibrotactile stimulation. Frames (a) through (c) present, respectively, a sample of the original MEG data, the whitened and the independent signals. Each tick corresponds to a time interval of 100 ms. The other frames show the field patterns associated with the first two independent components, and the brain sources corresponding to IC1 and IC2 superimposed onto vertical and horizontal MRI slices.

are approximately normal to the surface of the cortex. The white dots, based on the second independent component, correspond to the activation on the primary somatosensory area.

10.5.3   Segmenting Auditory Evoked Fields

In [39], using only averaged auditory evoked fields, we illustrated the decomposition capabilities of ICA in such setups. The stimuli consisted of 200 tone bursts that were presented to the subject's right ear, using 1. s interstimulus intervals. These bursts had a duration of 100. ms, and a frequency of 1. kHz. Figure 10.8 shows the 122 averages of the auditory evoked responses over the head. The insert, on the left, shows a sample enlargement of such averages, for an easier comparison with the results depicted in the next figures.

As in the study presented in the previous section, we can see from figure 10.9, that PCA is unable to resolve the complex brain response, whereas the ICA technique produces cleaner and sparser responses. For each component presented in figure 10.9 a left, top and right side views of the corresponding field pattern are shown. Note that the first principal component exhibits a clear dipole-like pattern over both the left and the right hemispheres, corroborating the idea of an unsuccessful segmentation of the evoked response. Subsequent principal components tend to have less and less structured patterns.

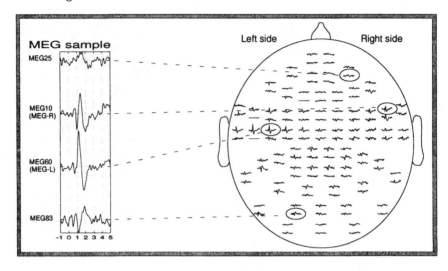

Fig. 10.8. Averaged auditory evoked responses to 200 tones, using a 122-channel device. Channels MEG10 and MEG60 are used in figure 10.9 as representatives of left-hemisphere and right-hemisphere MEG signals. Each tick in the MEG sample corresponds to 100 ms, going from 100 ms before stimulation onset to 500 ms after.

From the field patterns associated with the independent components we see that the evoked responses of the left hemisphere are isolated in IC1 and IC4. IC2 has stronger presence over the right hemisphere, and IC3 fails to show any clear field pattern structure. Furthermore, we can see that IC1 and IC2 correspond to responses typically labeled as N1m, with the characteristic latency of around 100 ms after the onset of the stimulation. The shorter latency of IC1, mainly reflecting activity contralateral to the stimulated ear agrees with the known information available for such studies.

Further support for the observations made above can be found in figure 10.10 where all four independent components are compared with one original left-hemisphere and one original right-hemisphere MEG signal, respectively. To solve the scaling indeterminations, each independent component is scaled to maximise the correlation with its corresponding original signal in both the left hemisphere and the right hemisphere. From the comparison of the independent components with the respective MEG signal, it is clear that IC1 and IC2 almost completely explain the N1m signals in the left and right hemispheres, respectively. IC4, exhibiting a longer latency (around 180 ms), fully explains the later responses in the left hemisphere. IC3 is a component that does not seem to be of significant importance, its scaled amplitude being much smaller than that of the other independent components.

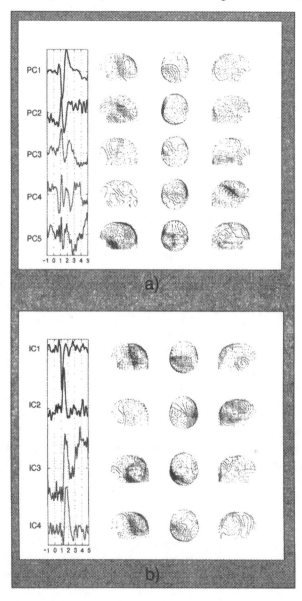

Fig. 10.9. a) Principal and b) independent components found from the auditory evoked field study. For each component, both the activation signal and three views of the corresponding field pattern are plotted.

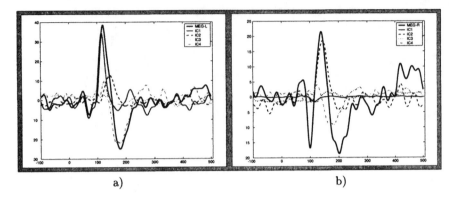

Fig. 10.10. The four ICs of figure 10.9 are plotted after scaling to a) one original left-hemisphere and b) one original right-hemisphere.

10.6   Conclusion

In this chapter we have shown examples of ICA in the analysis of biomagnetic brain signals. A special emphasis has been given to the justification of the ICA model for EEG and MEG signals.

The FastICA algorithm is suitable for extracting different types of artifact from EEG and MEG data, even in situations where these disturbances are smaller than the background brain activity.

ICA has been shown to be able to differentiate between somatosensory and auditory brain responses in the case of vibrotactile stimulation. In addition, the independent components, found with no other modeling assumption than their statistical independence, exhibit field patterns that agree with the conventional current dipole models. The equivalent current dipole sources corresponding to the independent components fell on the brain regions expected to be activated by the particular stimulus.

Finally, we have shown that the application of ICA to an averaged auditory evoked response nicely isolates the main response, with a latency of about 100 ms, from subsequent components. Furthermore, it discriminates between the ipsi- and contralateral principal responses in the brain. ICA may thus facilitate the understanding of the functioning of the human brain, as a finer mapping of the brain's responses may be achieved.

Although the results presented in this chapter are very promising, they are still preliminary, as we are just beginning to see the capabilities of ICA in the analysis of electro- and magnetoencephalographic recordings, as well as the possible limitations we may encounter when using ICA in biomedical applications. This work should, therefore, be seen as a door to the very promising set of possibilities opened to us by independent component analysis, rather than as a conclusion of such an approach.

References

1. *FastICA MATLAB package*. Available at
   http://www.cis.hut.fi/projects/ica/fastica.
2. S. Amari and A. Cichocki. Adaptive blind signal processing: Neural network approaches. *Proc. of the IEEE*, 86(10):2026–2048, 1998.
3. A. K. Barros, A. Mansour, and N. Ohnishi. Adaptive blind elimination of artifacts in ECG signals. In *Proc. of the 1998 Workshop on Independence & Artificial Neural Networks*, 1380–1386. Tenerife, Spain, 1998.
4. A. Bell and T. Sejnowski. An information-maximization approach to blind separation and blind deconvolution. *Neural Computation*, 7:1129–1159, 1995.
5. P. Berg and M. Scherg. A multiple source approach to the correction of eye artifacts. *Electroenceph. clin. Neurophysiol.*, 90:229–241, 1994.
6. S. Blanco, H. Garcia, R. Q. Quiroga, L. Romanelli, and O. A. Rosso. Stationarity of the EEG series. *IEEE Engineering in Medicine and Biology*, 395–399, 1995.
7. C. H. M. Brunia, J. Mcks, and M. V. den Berg-Lennsen. Correcting ocular artifacts: A comparison of several methods. *J. Psychophysiol.*, 3:1–50, 1989.
8. J.-F. Cardoso. Blind signal separation: statistical principles. *Proc. of the IEEE*, 86(10):2009–2025, 1998.
9. J.-F. Cardoso. High-order contrasts for independent component analysis. *Neural Computation*, 11:157 – 192, 1999.
10. P. Comon. Independent component analysis: a new concept? *Signal Processing*, 36:287–314, 1994.
11. M. Hämäläinen, R. Hari, R. Ilmoniemi, J. Knuutila, and O. V. Lounasmaa. Magnetoencephalography–theory, instrumentation, and applications to noninvasive studies of the working human brain. *Reviews of Modern Physics*, 65(2):413–497, 1993.
12. R. Hari. Magnetoencephalography as a tool of clinical neurophysiology. In E. Niedermeyer and F. L. da Silva, editors, *Electroencephalography: Basic principles, clinical applications, and related fields*, 1035–1061. Williams & Wilkins, Baltimore, 1993.
13. M. Huotilainen, R. J. Ilmoniemi, H. Tiitinen, J. Lavaikainen, K. Alho, M. Kajola, and R. Näätänen. The projection method in removing eye-blink artefacts from multichannel MEG measurements. In C. B. et al., editors, *Biomagnetism: Fundamental Research and Clinical Applications (Proc. 9th Int. Conf. Biomag.)*, pages 363–367. Elsevier, 1995.
14. J. Hurri, A. Hyvärinen, J. Karhunen, and E. Oja. Image feature extraction using independent component analysis. In *Pro. 1996 IEEE Nordic Signal Processing Symposium NORSIG'96*, 475–478, Espoo, Finland, 1996.
15. A. Hyvärinen. Survey on independent component analysis. *Neural Computing Surveys*, 2:94–128, 1999.
16. A. Hyvärinen and E. Oja. A fast fixed-point algorithm for independent component analysis. *Neural Computation*, 9(7):1483–1492, 1997.
17. A. Hyvärinen, J. Särelä, and R. Vig/'ario. Spikes and bumps: artefacts generated by independent component analysis with insufficient sample size. In *Proc. Int. Workshop on Independent Component Analysis and Blind Separation of Signals (ICA'99)*. Aussois, France, 1999.

18. A. Hyvrinen. Gaussian moments for noisy independent component analysis. *IEEE Signal Processing Letters*, **6**(6):145–147, 1999.
19. R. J. Ilmoniemi. *Biological effects of electric and magnetic fields*, **2**: 49–79. Academic Press, 1994.
20. B. W. Jervis, M. Coelho, and G. Morgan. Effect on EEG responses of removing ocular artifacts by proportional EOG subtraction. *Med. Biol. Eng. Comput.*, **27**:484–490, 1989.
21. B. D. Josephson. Possible new effects in superconductive tunneling. *Phys. Lett.*, **1**:251–253, 1962.
22. T.-P. Jung, C. Humphries, T.-W. Lee, S. Makeig, M. J. McKeown, V. Iragui, and T. Sejnowski. Extended ICA removes artifacts from electroencephalographic recordings. In *Neural Information Processing Systems 10*. MIT Press, Cambridge, MA, 1998.
23. C. Jutten and J. Hérault. Blind separation of sources, part I: an adaptive algorithm based on neuromimetic architecture. *Signal Processing*, 24:1–10, 1991.
24. J. Karhunen, E. Oja, L. Wang, R. Vigário, and J. Joutsensalo. A class of neural networks for independent component analysis. *IEEE Trans. Neural Networks*, **8**(3):486–504, 1997.
25. E. Kaukoranta, M. Hämäläinen, J. Sarvas, and R. Hari. Mixed and sensory nerve stimulations activate different cytoarchitectonic areas in the human primary somatosensory cortex SI. *Exp. Brain Res.*, **63**(1):60–66, 1986.
26. J. Le and A. Gevin. Method to reduce blur distortion from EEGs using a realistic head model. *IEEE Trans. Biomed. Eng.*, **40**:517–528, 1993.
27. S. Makeig, T.-P. Jung, A. Bell, D. Ghahremani, and T. Sejnowski. Blind separation of auditory event-related brain responses into independent components. *Proc. Natl. Acad. Sci. USA*, **94**:10979–10984, 1997.
28. M. McKeown, S. Makeig, S. Brown, T.-P. Jung, S. Kindermann, A. Bell, V. Iragui, and T. Sejnowski. Blind separation of functional magnetic resonance imaging (fMRI) data. *Human Brain Mapping*, **6**(5-6):368–372, 1998.
29. J. Mosher, P. Lewis, and R. Leahy. Multidipole modelling and localization from spatio-temporal MEG data. *IEEE Trans. Biomed. Eng.*, **39**:541–557, 1992.
30. E. Niedermeyer and F. Lopes da Silva, editors. *Electroencephalography. Basic principles, clinical applications, and related fields*. Williams & Wilkins, Baltimore, 1993.
31. B. A. Olshausen and D. J. Field. Emergence of simple-cell receptive field properties by learning a sparse code for natural images. *Nature*, **381**:607-609, 1996.
32. A. Papoulis. Probability, Random Variables, and Stochastic Processes. *Electrical & Electronic Engineering*. McGraw-Hill, Singapore, 3rd edition, 1991.
33. J. Sarvas. Basic mathematical and electromagnetic concepts of the biomagnetic inverse problems. *Phys. Med. Biol.*, **32**:11–22, 1987.
34. M. Scherg and D. von Cramon. Two bilateral sources of the late aep as identified by a spatio-temporal dipole model. *Electroenceph. clin. Neurophysiol.*, **62**:32 - 44, 1985.
35. A. W. Toga, R. S. J. Frackowiak, and J. C. Mazziotta, editors. *Brain Mapping Course (Satellite Course of the 4th International Conference on Functional Mapping of the Human Brain (HBM'98))*. Montreal, Quebec, Canada, 1998.
36. K. Torkkola. Blind separation for audio signals: Are we there yet? In *Proc. Int. Workshop on Independent Component Analysis and Blind Separation of Signals (ICA'99)*. Aussois, France, 1999.

37. R. Vigário. Extraction of ocular artifacts from EEG using independent component analysis. *Electroenceph. clin. Neurophysiol.*, **103**:395–404, 1997.
38. R. Vigário. *Independent Component Approach to the Analysis of EEG and MEG Recordings*. PhD thesis, Helsinki University of Technology, 1999.
39. R. Vigário, V. Jousmäki, M. Hämäläinen, R. Hari, and E. Oja. Independent component analysis for identification of artifacts in magnetoencephalographic recordings. In M. I. Jordan, M. J. Kearns, and S. A. Solla, editors, *Neural Information Processing Systems 10*. MIT Press, Cambridge, MA, 1998. MIT Press.
40. R. Vigário, J. Särelä, V. Jousmäki, M. Hämäläinen, and E. Oja. Independent component approach to the analysis of EEG and MEG recordings. *IEEE Trans. Biomed. Eng.*, 2000. (To appear).
41. R. Vigário, J. Särelä, V. Jousmäki, and E. Oja. Independent component analysis in decomposition of auditory and somatosensory evoked fields. In *Proc. Int. Workshop on Independent Component Analysis and Blind Separation of Signals (ICA'99)*. Aussois, France, 1999.
42. R. Vigário, J. Särelä, and E. Oja. Independent component analysis in wave decomposition of auditory evoked fields. In *Proc. Int. Conf. on Artificial Neural Networks (ICANN'98)*, Skövde. Sweden, 1998.
43. J. O. Wisbeck, A. K. Barros, and R. G. Ojeda. Application of ICA in the separation of breathing artifacts in ECG signals. In *Proc. of the Conf. on Neural Information Processing (ICONIP'98)*, 211–214. Kitakyushu, Japan, 1998.

11 ICA on Noisy Data: A Factor Analysis Approach

Shiro Ikeda

11.1 Introduction

The basic problem of ICA is defined for the noiseless case, where the sources and observations have the following linear relation,

$$x = As \tag{11.1}$$
$$x \in R^n, \quad s \in R^m, \quad A \in R^{n \times m}$$

The assumptions of an ICA problem are that the mean value of each component of s is 0, mutually independent and drawn from different probability distributions which are not Gaussian expect for one. We also restrict m to be smaller than or equal to n in this chapter for the existence of linear solution.

The aim of ICA is to estimate a matrix W which satisfies the following equation,

$$WA = PD \tag{11.2}$$
$$W \in R^{m \times n}, \quad P \in R^{m \times m}, \quad D \in R^{m \times m}$$

Here, P is a permutation matrix which has a single entry of one in each row and column and D is a diagonal matrix. This simple problem is solved in the framework of a semi-parametric approach and gives a lot of interesting theoretical and practical results.

However, when we apply ICA to real-world problems, the situation is different from the above ideal case. In many cases, we cannot avoid the effect of noise and the number of the sources m is not known. These two problems are related to each other since, if there is no noise, m can be determined directly from the data. For example, in the case of biological data such as EEG or MEG, the number n of the measurements is large and sometimes around 200, but we believe that the number of the sources is not so large in a macroscopic viewpoint and the noise is very great. Therefore, equation 11.1 is not enough to describe the problem. It is pointed out that, especially when the number of the sources is small, one cannot have a good solution generally [4].

In this chapter, we discuss the case where there are additive noises in observations as,

$$x = As + \epsilon \tag{11.3}$$

Here, ϵ is an n-dimensional real valued noise term and we assume that components ϵ_i of ϵ are mutually uncorrelated. This is the case in MEG data.

The form of equation 11.3 is equivalent to the factor analysis model when the noises and the sources are from Gaussian distributions. A method to solve this problem in the framework of maximum likelihood estimation by modeling the source distributions in terms of Gaussian mixtures has been proposed [2]. It is one of the extensions of basic ICA problem, but it is difficult to apply it to biological data when the number of inputs is very large.

We solve the problem with a semi-parametric approach. The idea is to use factor analysis for the preprocessing of data. This is equivalent to replacing PCA, which is widely used as the preprocessing technique for ICA , by factor analysis. We estimate the power of the noises and also the number of the sources by factor analysis. After the preprocessing, we use one of the ICA techniques to estimate the separation matrix. In the following sections, we describe the factor analysis approach, its relation to ICA, and experimental results.

11.2 Factor Analysis and ICA

11.2.1 Factor Analysis

In factor analysis, we suppose the case where real-valued n-dimensional observed data x are modeled as,

$$x = \mu + Af + \epsilon \tag{11.4}$$

$$x, \mu, \epsilon \in R^n, \quad f \in R^m, \quad A \in R^{n \times m}$$

The mean of x is given by μ which is assumed to be 0 in this chapter. Without μ, this equation becomes similar to equation 11.3. The assumptions in factor analysis are that: a) f is jointly normally distributed as $f \sim N(0, I_m)$ where I_m is the $R^{m \times m}$ identity matrix, b) ϵ is normally distributed as $\epsilon \sim N(0, \Sigma)$, where Σ is a diagonal matrix, and c) f and ϵ are mutually independent.

The goal of factor analysis is to estimate m, A (the factor loading matrix), and Σ (the unique variance matrix) using the second order statistics of the observation x, which is defined as $\overline{xx^T}$. When m is given, there are various estimation methods for A and Σ. The major ones are the unweighted least squares method (ULS) and the maximum likelihood estimation (MLE).

Both of them are summarized below in terms of related loss functions. Suppose we have a data set as $\{x_i\}$ $(i = 1, \ldots, N)$ and let C be the covariance matrix of observed data x, $(C = \sum x_i x_i^T / N)$. The estimate of ULS $(\hat{A}, \hat{\Sigma})_{\text{ULS}}$

and that of MLE $(\hat{A}, \hat{\Sigma})_{\mathrm{MLE}}$ are defined as,

$$(\hat{A}, \hat{\Sigma})_{\mathrm{ULS}} = \operatorname*{argmin}_{A, \Sigma} \ \mathrm{tr}\,(C - (AA^T + \Sigma))^2 \tag{11.5}$$

$$(\hat{A}, \hat{\Sigma})_{\mathrm{MLE}} = \operatorname*{argmax}_{A, \Sigma} L(A, \Sigma) \tag{11.6}$$

$$L(A, \Sigma) = -\frac{1}{2} \left\{ \mathrm{tr}\,\left(C(\Sigma + AA^T)^{-1}\right) \right. \tag{11.7}$$
$$\left. + \log(\det(\Sigma + AA^T)) + n \log 2\pi \right\}$$

To solve the equations, we can use the gradient descent algorithm or Gauss-Newton method. Also the Expectation Maximization EM algorithm can be applied to MLE. But we don't go into the detail of them in this chapter.

The next problem is to estimate the number of factors. There are many approaches to estimating m. Some of them are based on the eigenvalues of the covariance matrix. There are also some other well-known approaches which are based on the model selection approach with a kind of information criterion such as Akaike Information Criteria (AIC) and Minimum Description Length (MDL). We use MDL for the estimation of m in the framework of model selection. Since MDL is based on MLE, we use MLE for the estimation of A and Σ as shown in equation 11.6.

MDL is defined as follows,

$$\mathrm{MDL} = -L(\hat{A}, \hat{\Sigma}) + \frac{\log N}{N} \times \text{the number of free parameters} \tag{11.8}$$

We have to know the number of the free parameters in the model to calculate MDL. The factor analysis model has $n(m + 1)$ parameters in A and Σ but for an arbitrary $m \times m$ unitary matrix, T_m, $B = AT_m$ has the same likelihood function as A. This shows that A is determined to within ambiguities of the rotation. The degree of freedom of T_m are $m(m-1)/2$ and we should subtract $m(m-1)/2$ from $n(m+1)$; the number of free parameters is $n(m+1) - m(m-1)/2$ and

$$\mathrm{MDL} = -L(\hat{A}, \hat{\Sigma}) + \frac{\log N}{N}\left(n(m + 1) - \frac{m(m - 1)}{2}\right) \tag{11.9}$$

We also have a constraint for the existence of the estimates. C has $n(n+1)/2$ degrees of freedom and we need the constraint that $n(n + 1)/2 \geq n(m + 1) - m(m - 1)/2$. By taking one more condition $m < n$ into account, we have the following bound,

$$m \leq \frac{1}{2} \left\{ 2n + 1 - \sqrt{8n + 1} \right\} \tag{11.10}$$

This is a necessary condition for A to be estimable (Ledermann [5]). Anderson and Rubin also discussed sufficient conditions [1].

11.2.2   Factor Analysis in Preprocessing

In many cases, ICA algorithms are separated into two parts. One preprocesses the data such that they become uncorrelated. This part is usually called sphering or whitening. After this preprocessing, we search for the remaining rotation matrix using some algorithm.

For sphering, the basic approach is to use PCA. This is based on the second order statistics and is only effective for noiseless cases. Let $P = C^{1/2}$, where $C = PP^T$. By letting,

$$x' = P^{-1}x$$

it is clear that $\sum x'x'^T/N = I_n$. This is equivalent to assuming that the power of each source signal, s_i, is 1. We can then remove the ambiguity of amplitude.

When the data are noisy, we still make the assumption that the sources are uncorrelated and that each power is 1, that is $E[ss^T] = I_m$. But the matrix $P = C^{1/2}$ does not give preprocessing such that the remaining part is an orthogonal matrix. We can instead use factor analysis. As the result of factor analysis, we can estimate $\hat{A}$ and $\hat{\Sigma}$. Let Q be the pseudo-inverse of $\hat{A}$ where $\hat{A}Q\hat{A} = \hat{A}$ holds. Q is an $m \times n$ real valued matrix.

After transforming the data as,

$$z = Qx$$

z becomes the sphered data. However, its covariance matrix is not the identity matrix. Let the true parameters be A^* and Σ^*, and let Q^* be the pseudo-inverse of A^*. Then,

$$\frac{1}{N}\sum z^*z^{*T} = I_m + Q^*\Sigma^*Q^{*T}$$

This result shows that we can make the part of observation x due to the sources uncorrelated, but not the noises at the same time.

11.2.3   ICA as Determining the Rotation Matrix

After we preprocess the data by Q, they are uncorrelated except for the part due to the noise. What is left to determine is the rotation matrix. In factor analysis, this is also a big problem. But here, instead of adopting a factor analysis approach to the problem, we can use approaches by ICA.

In section 11.2.1, we assumed that s and ϵ are normally distributed. We now break that assumption. We still assume that ϵ are normally distributed, but s is not normally distributed and each component is independent. We can use an ICA algorithm now.

The ICA algorithm should not be affected by the second order statistics since, even if data are preprocessed by factor analysis, data z still has second order correlations. Therefore, an algorithm based on higher order correlations is preferable here. We used the JADE algorithm by Cardoso [3] which is based on the 4th order cumulant. In the next section, we show some experimental results.

11.3   Experiment with Synthesized Data

First, we use speech data which are recorded separately and mixed on the computer.

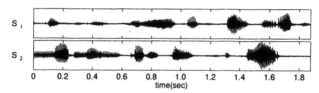

Fig. 11.1. Source sound signals: speech signals recorded separately with 16.kHz sampling rate. Data are 1.875.sec length and we have 30,000 data points.

The source data are shown in figure 11.1. We have two sources and we suppose that there are seven sensors and noises are added on each sensors separately as,

$$x = A^*s + \epsilon \tag{11.11}$$

$$x, \epsilon \in R^7, \quad s \in R^2, \quad A^* \in R^{7\times 2}, \quad \epsilon \sim N(0, \Sigma^*)$$

We set the power of the source signals to be equal to that of the noises in the observations, which means,

$$\text{tr}\left(A^* \overline{ss^T} A^{*T}\right) = \text{tr}\left(A^* A^{*T}\right) = \text{tr}\,\Sigma^* \tag{11.12}$$

The observations x are shown in figure 11.2. The data are noisy and it is impossible to use the usual sphering method. Also, the number of the sources is assumed to be unknown.

First, we apply factor analysis on the data and estimate the mixing process, noise covariance, and the number of sources. In this case, $m \le 3.7251\ldots$ should be satisfied from equation 11.10, and the candidates for the number of the sources are $1, 2$ and 3.

We used MDL to estimate m. For comparison, we also show AIC for each candidate in Table 11.1.

Both MDL and AIC selected two as the source number. But AIC gives a very small difference for $m = 2$ and 3. This is the reason why we used MDL for the other problems in this chapter. We selected the source number as 2

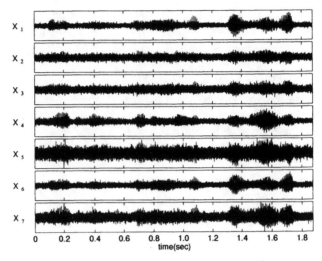

Fig. 11.2. The seven observed mixed and noisy input signals.

Table 11.1. MDL and AIC for the candidates.

# of sources	1	2	3
MDL	4.0870	3.8911	3.8928
AIC	4.0826	3.8849	3.8850

and obtained the pseudo-inverse of $\hat{A}$ as $Q \in R^{2 \times 7}$. We transform the inputs with this Q as,

$$z = Qx$$

The resultant z is shown in two ways in figure 11.3.

We can see that the sources are orthogonal, that is, uncorrelated to each other. Thus, preprocessing was almost complete. The remaining problem is to estimate the rotation matrix. We estimated the rotation matrix by the JADE algorithm and estimated the demixing matrix $W \in R^{2 \times 2}$. This is equivalent to making the fourth order statistics uncorrelated. Finally, we linearly transformed the signals as,

$$y = WQx$$

Here, y is of two dimensions and it includes noises. However, from figure 11.4, one can see the original sources are recovered very well.

In figure 11.5, we show estimated covariance matrixes and true covariance matrixes as 2-dimensional ellipsoids. From the figure we show that our

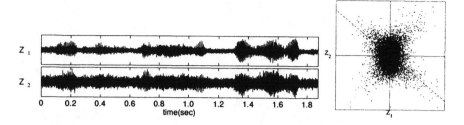

Fig. 11.3. The preprocessed data after having been sphered.

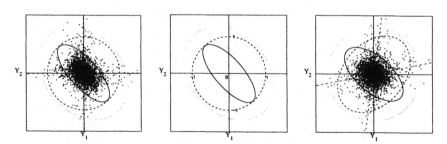

Fig. 11.4. The data after having been sphered and finally rotated.

method gives a good estimation and, as a result, we can estimate the separation matrix.

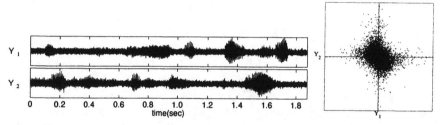

Fig. 11.5. Estimated covariance matrixes (left figure), true covariance matrixes (center figure) and result of PCA+JADE algorithm (right figure): solid ellipses show the covariance matrix of the noise term which is $WQ\Sigma Q^T W^T$, dashed ellipses show the covariance matrix of the sources which are unit matrix I_m, dotted ellipses show the covariance matrix of the separated data which is $I_m + WQ\Sigma Q^T W^T$ and dashed dot lines show the axes of the sources which are x and y axes in the center figure. The data points shown in the left graph were rescaled to match the size.

In this experiment, we know the true mixing matrix A^*, therefore we can calculate the cross-talk as the ratio of diagonal and off-diagonal components of the matrix WQA^*. In $y_1(t)$, the cross-talk is 0.92% and in $y_2(t)$, it is 0.57%.

We also tried PCA on the data and selected a 2-dimensional subspace by PCA based on the eigenvalues. The data was based on the eigenvalues of the covariance matrixes. After compressing the data, we applied JADE on the compressed data and estimated a rotation matrix. The result is shown in figure 11.5. In the case of PCA+JADE, we cannot make the sources orthogonal to each other by the preprocessing and the separation matrix is not estimated correctly.

11.4   MEG Data Analysis

We also applied our method to MEG data. Before going into the detail of the experiment, we discuss the characteristics of the MEG data.

MEG measures the magnetic field caused by brain activity with many (50~200) coils placed around the brain. Because the change in the magnetic field is directly connected to the nerve activities, MEG can measure the brain activities without any delay. Also the sampling rate is around 1.kHz which is much higher than other techniques such as MRI. We can also estimate the location of the sources using some techniques by solving the inverse problem, and the resolution can be in the order of a few mm. Therefore, MEG is a technique which measures the brain activity with high time and spatial resolutions without giving any damage to the brain.

Since the magnetic field caused by the brain activity is extremely small ($\sim 10^{-14}$ T), we need special device called a Super-conducting Quantum Interference Device (SQUID). The device can detect the brain signal, but the signal contains a lot of environmental noises. We can categorize the noises into two major categories. One is called the artifacts and the other is the quantum mechanical noise. The artifacts include noises from an electric power supply, the earth's magnetism, the heart beat, breathing and the brain activity in which we are not interested. These artifacts affect all the sensors simultaneously. On the other hand, the quantum mechanical noise originates from the SQUID itself. SQUID measures the magnetic field in liquid helium. At this low temperature, sensors cannot avoid having quantum mechanical noise which is white and independent from each other. The main technique used so far to reduce the noise is averaging. The experimenter usually repeats the same experiment 100 to 200 times and then averages the recorded response. But the averaged data still contains a lot of noise, which is still harmful for the data analysis. After averaging the data, we modeled the data as,

$$x(t) \ = As(t) + \epsilon(t) \tag{11.13}$$

$s(t)$: sources and artifacts

$\epsilon(t) \sim N(0, \Sigma)$: quantum mechanical noise

We then applied our algorithm to MEG data. In the following sections, we show the results of phantom MEG data and real MEG data.

11.4.1   Experiment with Phantom Data

First, we applied our algorithm to phantom data. The phantom was designed to be roughly the same size as the brain and there was a small platinum electrode inside. The signal was a triangle wave with frequency 20.Hz and we averaged the data for 100 trials. Figure 11.6 shows some of the signals on the sensors. There are 126 active sensors and we are showing only three of them. We applied our algorithm for selecting the number of the sources and

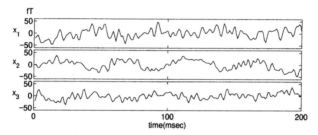

Fig. 11.6. The sensor outputs obtained by using the phantom data.

the separation matrix. We preprocessed the data with factor analysis and estimated the number of the sources by MDL. After the rotation matrix was estimated by the JADE algorithm, we applied the pseudo-inverse matrix to the preprocessed data. The estimated independent signals are shown in figure 11.7.

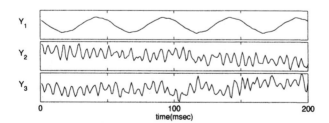

Fig. 11.7. Traces of the eventual estimated independent components

In this experiment, we know the input to the electrode is Y_1 in figure 11.7. After selecting the source, we want to reconstruct the signal. This is possible because we have estimated the matrix A. Each column of $\hat{A}$ corresponds to the coefficients of how each of the components contributes to the inputs of the sensors. Let $\hat{A} = (\hat{a}_1, \ldots, \hat{a}_m)$, where $\hat{a}_i$ is an n-dimensional column vector. In this case, we can reconstruct the data with $x'(t) = a_1 y_1(t)$. The recovered signals are shown in figure 11.8. These are the recovered signals on the sensors in figure 11.6. We can see that the noise is reduced by comparing these two figures.

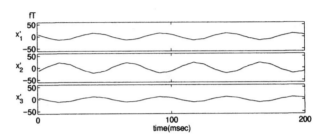

Fig. 11.8. The final recovered independent components using the proposed method.

Figure 11.9 shows the estimated strength of each source and the noise on the sensors. The strength of the source is estimated as the component of matrix $\hat{A}$. Also the noise is estimated as the square root of the diagonal component of the matrix $\hat{\Sigma}$. From the graph, we can see that some sensors contain much more noise and artifacts than signals, even after averaging over 100 trials. This result shows that we cannot always trust the intensities of the signals on the sensors. When we estimate the location of the signal using some algorithms, we have to know the intensity ratio of the signals on all the sensors, but the result shows that artifacts and noises are sometimes harmful.

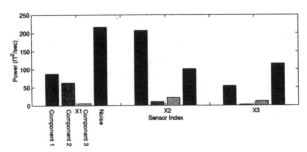

Fig. 11.9. Estimated strength of each source and the noise on the sensors.

11.4.2   Experiment with Real Brain Data

We also applied our algorithm to the data of the brain activity evoked by visual stimulation. One of the difficulties in evaluating brain data is that we don't know the true signal. This is a big problem in knowing how well our algorithm is working.

Searching for the independent components we can have through the study of MEG data, we can summarize the expected result of the ICA applied to MEG data as follows.

1. Separating artifacts and brain signals.
2. Separating the independent brain activities coming from different parts of the brain.

For the first part, we believe this is possible because the artifacts and the brain signals would be independent. But for the second part, we don't know if the brain activity which is coming from different parts are independent from each other. It is more intuitive to think that they might be dependent. This is a difficult problem that forces us to go further in generalization of the ICA framework. Before going on to discuss those problems, we show the separated

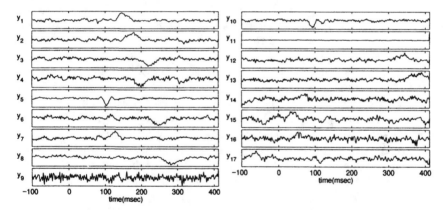

Fig. 11.10. Separated visual evoked response signals. The independent components are aligned along the descent order of their averaged powers on the MEG sensors. 0 msec in the time axis is the trigger of the visual stimulation.

independent components obtained by our method in figure 11.10. The same kind of visual stimulations are given to a subject. The data are recorded by 114 sensors in this case. The duration of recording is from 100.msec before the stimulation until 412.msec after the stimulation with 1.kHz sampling rate. The same procedure was applied to one subject 100 times and we averaged the data. Some outputs from the sensors are shown in figure 11.11. It is observed that the averaging reduces the noise but a lot of noise still remains

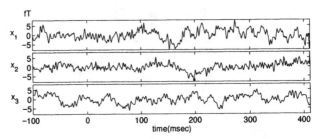

Fig. 11.11. The MEG data: showing the averaged data of three sensors.

in the data. We applied our method to the data and selected 17 independent components by MDL in this experiment. The independent components are shown in figure 11.10. We also applied the method to the data from a further four different subjects. In all the cases, the number of sources selected is roughly the same (from 16 to 19).

Based on the results in figure 11.10, we have to separate the sources from the artifacts in which we are not interested. For example, we can see that $y_9(t)$ mainly contains high frequency (mainly 180.Hz) signal which seems to be a noise component from electric power supply. Similarly $y_1 1$ has a very large value at the very end of the recording which seems to correspond to some kind of mechanical noise of the recording. But for the other 15 components, we cannot know if they are due to brain sources or not. Fortunately, this experiment is designed for studying evoked responses by visual stimuli and we are not interested in the components which have some power before stimuli were shown to the subject. Therefore, we defined a threshold of power such that if a signal has some power before the stimulation we regarded the signal as an artifact. Also we added one more criterion that if the estimated averaged power of an independent component on all the sensors is smaller than a threshold (in this case, $900[\text{fT}^2/\text{sec}]$), we also assumed the signal to be an artifact. In this case, the selected remaining sources are $y_1(t)$, $y_2(t)$, $y_3(t)$, $y_5(t)$, $y_6(t)$, $y_8(t)$, and $y_{10}(t)$.

After picking those sources up, we put them back into the original sensor signal space and the result is shown in figure 11.12. The noises are removed and the data are clear.

We can see the cleaned outputs of the sensors from this result, but we further want to know the relationship between the independent components and brain activities more directly. There are many ways of seeing the relationships visually. One popular method is the dipole estimation. In the dipole estimation, we describe the brain activities by dipoles and we have to specify the number of the dipoles. Since the number of the dipoles is not clear in the case of the real brain model, we did not employ dipole estimation, but we implemented the spatial filter (SF) technique [6]. SFs are virtual sensors which are located on the brain. We want to know the current flows on those

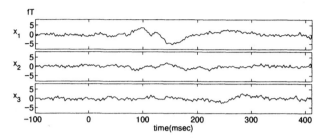

Fig. 11.12. The recovered MEG data after the artifacts are removed.

virtual sensors which describe the MEG observations well. This is an inverse problem and can be represented as a linear mapping from the MEG sensors to SFs.

One of the characteristics of SFs is that we can put virtual sensors at any place we want, so that we can estimate the activity of any place on the brain virtually by a linear projection. But, from the nature of this technique, the sensors at the boundary are not reliable [6]. In our experiment, the part of the brain we are interested in is the visual cortex and we put 21×21 SFs on a half sphere, whose center is located at V1. Figure 11.13 shows the output of the SFs before and after the method. Since the original data includes a lot of noise, we cannot know if there is useful information in the SF, but after removing the noises, we can see the brain activity around the visual cortex clearly.

11.5   Conclusion

In this chapter, we proposed a new combination of methods from different backgrounds to analyse biological data. The data includes strong noises, which is true of most biological data. We applied the algorithm to MEG data, and have shown the approach is effective. We can estimate the number of sources, and the power of the noises on each sensor independent from each other. This is one of the serious problems which has not been well treated in conventional ICA approaches and this chapter gives one effective approach. Attias proposed a method to solve this problem: it gives the source distribution in the form of a normal mixture and everything is solved by using the parameters in MLE. But we have proposed a different method based on the semi-parametric approach which is an attractive part of ICA.

This approach is a natural extension of the standard ICA approach which uses PCA first and then higher order statistics. However, there still remain a lot of open problems. In factor analysis, there are a lot of methods for

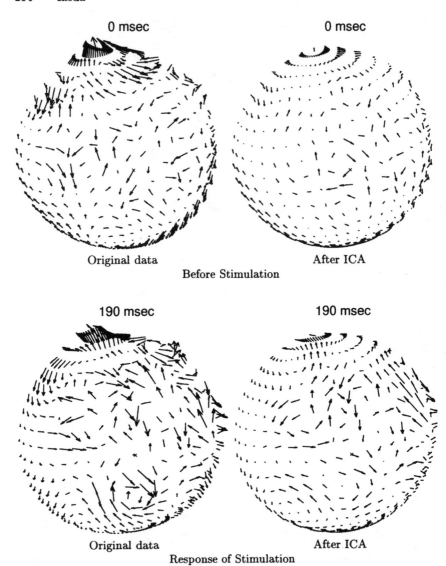

Fig. 11.13. Result of the approach applied to MEG data: Outputs of SFs

estimating the parameters and the number of the sources, and each method has its own characteristics. We applied MLE for estimation and MDL for estimating the number of the sources. But there are different combinations and there may be a method which suits particular problems better. The same may be true for the ICA algorithms. We used JADE but there may be a better algorithm. Another problem is the noise distribution. We assumed Gaussian distributions but if we can have a better model for the noise, the algorithm will be improved further.

We can also check the algorithm from the viewpoint of factor analysis. How to determine the rotation is one of the traditional problems in factor analysis. There are a lot of algorithms proposed for this purpose, but it is not common to use higher-order statistics. Therefore our approach also gives a new pathway for factor analysis, too.

11.6   Acknowledgements

The author is grateful to Shun-ichi Amari, Keisuke Toyama, and Noboru Murata for their comments and suggestions for the research. The author also thank Shigeki Kajihara and Shimadzu Inc. for the MEG data.

References

1. T. W, Anderson, and H. Rubin. Statistical inference in factor analysis, *Proceedings of the third Berkeley Symposium on Mathematical Statistics and Probability*,**5**: 111 - 150, 1956.
2. H. Attias, Independent Factor Analysis. *Neural Computation*, **11**:(4), 803-851, 1999.
3. Jean-Franccois Cardoso and Antoine Souloumiac, Blind beamforming for non Gaussian signals., *IEE-Proceedings-F*,**140**:(6), 362 - 370, 1993.
4. A. Hyvärinen, J. Särelä and R. Vigário . Spikes and bumps: Artifacts generated by independent component analysis with insufficient sample size. *Proceedings of International Workshop on Independent Component Analysis and Blind Signal Separation (ICA'99)*, 425 - 429, 1999.
5. W. Ledermann,On the rank of the reduced correlational matrix in multiple-factor analysis. *Psychometrika*, **2**: 85 - 93, 1937.
6. K. Toyama and K. Yoshikawa and Y. Yoshida and Y. Kondo and S. Tomita and Y. Takanashi and Y. Ejima and S. Yoshizawa, A new method for magnetoencephalography: A three dimensional magnetometer-spatial filter system., *Neuroscience*, **91**:(2), 405 - 415, 1999.

12 Analysis of Optical Imaging Data Using Weak Models and ICA

John Porrill, James V. Stone, Jason Berwick, John Mayhew
and Peter Coffey

12.1 Introduction

This chapter will present techniques for the recovery of component spatial and temporal modes from spatio-temporal data sets, in particular from medical imaging data such as that obtained by functional magnetic resonance imaging functional magnetic resonance imaging (fMRI) and optical imaging optical imaging (OI) of brain activity. These techniques were developed in order to help address some current issues involving the nature of the haemodynamic response to neural activity (for more details of this research and further references see [7]).

Current analysis techniques can be divided, roughly speaking, into two classes: data-driven and model-driven. An example of a data driven technique is principal component analysis (PCA). This requires no prior knowledge about the form of the components but physical interpretation of the decomposition is difficult and there is no obvious way to test for statistical significance. In contrast, model-driven techniques such as general linear model analysis require that constrained models of the expected behaviour be prescribed in advance but have the very significant advantage that they allow the use of standard statistical tests.

In our application, it is important that the components we recover correspond to underlying physical processes. Since current biophysical knowledge of these processes is limited, the decomposition must be based, as far as is possible, on generic principles rather than on highly constrained models. We therefore define an intermediate class of *weak temporal models*, designed to utilise incomplete prior information about temporal responses (available, for example, from experimental protocols) without having to specify the expected behaviour in detail.

We will show how to define a *weak causal model* for the response to a single stimulus and a *weak periodic model* for the response to a periodic (e.g. box-car) stimulus, and discuss how the statistical significance of temporal components recovered using such models may be assessed.

The weak models we define here prescribe an optimal temporal response but leave the problem of recovering the associated spatial activity map underconstrained. We show that this ill-posed problem can be regularised effectively using the entropy measure underlying spatial independent component analysis. independent component analysis We will argue that this regulariser can be regarded as implementing a spatial prior based on knowledge of the spatial statistics of various classes of activity map.

We will illustrate the techniques described in this paper by applying them to a data set obtained by direct optical imaging of a rat barrel cortex before, during, and after whisker stimulation. The images have been preprocessed to estimate optical absorption at the illumination wavelength of 577 nm and averaged over 30 trials to reduce the contribution of background vascular processes (in particular a dominant 0.1 Hz vasomotion signal). Each data set contains 12 s of data sampled at 15 Hz. Whisker stimulation began 8 s after image aquisition started and continued for 1 s, image aquisition then continued for a further 3 s. The final data set consists of 180 128 × 96 images. For more details of the experimental procedure, a description of the non-linear spectral analysis method used to estimate physical parameters from multi-spectral data, and an assessment of the performance of various analysis methods, see [7].

This data set is representative in that it contains a number of confounds, such as residual background effects, movement artifacts, and camera noise, as well as the responses to stimulation which interest us.

12.2   Linear Component Analysis

We will represent a spatio-temporal data set by an $m \times n$ matrix X where m is the number of spatial pixels and n is the number of temporal slices.

To simplify the discussion both row and column means are assumed to be removed from X (in our problem non-zero means correspond to an un-varying background image or an overall variation in illumination; both are uninteresting from the point of view of this analysis).

A *linear decomposition* of X then has the form

$$X = \sum_i s_i \otimes t_i \tag{12.1}$$

and is equivalent to a matrix factorisation

$$X = ST^t \tag{12.2}$$

where pixel vectors s_i for spatial modes are stored in the columns of $S = (s_1 | \ldots | s_k)$ and and their associated time variations t_i in the columns of $T = (t_1 | \ldots | t_k)$.

Although infinitely many such factorisations of X are possible we hope to find a factorisation in which modes (s_i, t_i) correspond to underlying physical processes. This is clearly possible in principle if these processes are linear. For the applications with which we are concerned, approximate linearity is a reasonable working hypothesis given the small amplitude of the effects under consideration.

Matrix factorisations are usually algebraic, defined by placing algebraic constraints such as orthonormality or triangularity on the factors. The size of our data sets allows us to consider an alternative class of *stochastic factorisations* where factors are characterised in terms of the statistical properties of the populations from which their columns are drawn. In fact the first factorisation we consider, singular value decomposition, belongs to both classes, being defined by algebraic constraints which have a natural statistical interpretation.

12.3 Singular Value Decomposition

The singular value decomposition (SVD) of X

$$X = UDV^t \tag{12.3}$$

is defined by the constraints

$$U^tU = I, \qquad V^tV = I, \qquad D \text{ diagonal} \tag{12.4}$$

These algebraic constraints can be interpreted statistically: they can be derived from the hypothesis that spatial and temporal modes are pairwise uncorrelated. In the next section, we apply SVD to our data set and discuss the range of validity of this hypothesis.

Our main application of SVD will be to reduce the dimensionality of the factorisation problem and we turn to this problem now. The minimum possible number of modes k in an exact linear decomposition is given by the rank of the matrix X. For generic data sets, this number will usually be large (the minimum of the matrix row and column sizes m and n). This number can however be reduced if we are satisfied with an approximate factorisation

$$X = PQ^t + E \tag{12.5}$$

For such an approximate factorisation of rank k it is known that the factorisation error $||E||_2$ is minimised when

$$P = U_k \qquad Q = V_kD_k \tag{12.6}$$

where U_k, V_k retain only those columns of U, V (the principal spatial and temporal components) corresponding to the largest k singular values in D.

A suitable number of modes k to be retained can be estimated by various methods. The most satisfactory criterion is to make the proportion of variance explained by the components retained close to unity. If this does not reduce dimensionality sufficiently, a sharp transition in the amount of variance explained per component may indicate the point at which sensor noise begins to dominate. This choice can be confirmed by visual inspection of the spatial and temporal modes, setting the cut-off when they become too unstructured to be useful in describing biological processes.

Suppose an approximate rank k factorisation

$$X \approx PQ^t \tag{12.7}$$

is available. Then the family of rank k factorisations

$$X \approx ST^t \tag{12.8}$$

with the same error matrix can be parameterised using spatial and temporal unmixing matrices W_S and W_T:

$$S = PW_S \qquad T = QW_T \tag{12.9}$$

subject to the constraint

$$W_T = W_S^{-t} \tag{12.10}$$

(we use the notation A^{-t} throughout to denote the inverse transpose $(A^t)^{-1}$). The approximate factorisation problem is then reduced to the recovery of the k^2 free parameters in either W_S or W_T.

12.3.1   SVD Applied to OI Data Set

Figure 12.1 shows the variance explained plot for the principal components of the OI data set.

The clear shoulder on this curve suggests an appropriate cut-off point for dimension reduction. In the following analyses we use $k = 6$ since this is also the point at which the spatial modes become unstructured. Analyses not presented here have used much larger values without affecting the conclusions significantly. Figure 12.2 shows the first six principal spatial and temporal components. The analysis has clearly revealed some interesting stucture and activity. For example the first spatial principal component seems to correspond mostly to vascular activity and the associated temporal response is consistent with this being the remnant after averaging of the 0.1 Hz vasomotion signal.

The second spatial component shows a compact region of spatial activity in an area which was subsequently identified histologically with the whisker barrel region. The temporal behaviour of this mode is not a simple response to the stimulus (applied in the period bounded by vertical dotted lines), although it does seem to have been strongly affected by it.

The fourth spatial component looks very like a derivative image and this, taken together with its periodic behaviour in time, strongly suggests that it is a motion artifact, possibly due to breathing or heartbeat. Though it is tempting to identify principal components with underlying physical components in

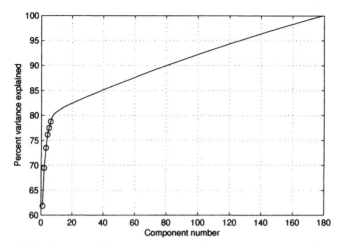

Fig. 12.1. The plot shows the cumulative percentage variance explained by the principal components of the OI data set. The six circled values before the curve shoulder correspond to components which will be retained in subsequent analyses.

this way it is important to realise that this process is subject to important limitations. For example the second temporal principal component in figure 12.2 cannot unambiguously be interpreted as the temporal response of the barrel since it is necessarily orthogonal to the residual vasomotion recovered in the first component and so is determined in part by what is effectively system noise.

Interpretation of principal components as underlying physical modes is clearly unsafe whenever the time courses of these modes are likely to be correlated. In situations where there are multiple biophysical responses to the same stimulus this can hardly be avoided.

Given these problems of interpretation it is interesting to consider why the motion artifact is well determined by SVD. In fact this mode satisfies the statistical constraints underlying SVD: it is approximately uncorrelated, both in space and time, with other physical modes. This is because a small amplitude motion has a spatial map which is a derivative image, and so is approximately orthogonal to generic smooth images, and high frequency periodic temporal behaviour is approximately orthogonal to generic smooth temporal modes. This component has been correctly recovered because it has been correctly modelled!

12.4   Independent Component Analysis

Independent component analysis (ICA, see [2]) is closely related to principal component analysis but attempts to recover components which are not just

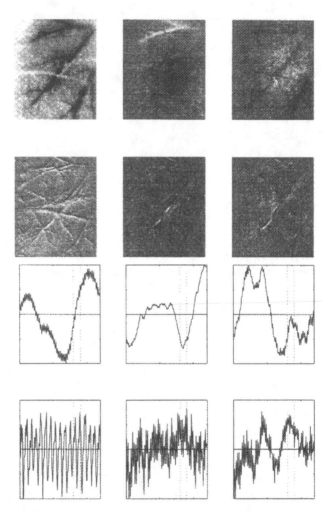

Fig. 12.2. Results of applying SVD to the OI data set. The images in the top panel show the six principal spatial modes (counting left to right, top to bottom) and the plots below show their associated temporal variations (with dotted lines bounding the period of whisker stimulation).

uncorrelated but statistically independent. When applied to spatio-temporal data sets it can take two forms, spatial ICA and temporal ICA (sICA and tICA). Recently sICA has been applied to the analysis of fMRI data sets with some success [6].

The sICA factorisation can be regarded as implementing a statistical model in which the data set is an unknown combination of independent spatial modes s_i whose pixels are independently sampled from known pdfs p_i. No prior constraints are placed on the temporal modes. This sidesteps the orthogonality problem we discussed above the unrealistic constraint that temporal modes be uncorrelated is not needed since sICA enforces stronger independence constraints on spatial modes. Note that since uncorrelated Gaussian random variables are necessarily statistically independent, sICA can only work when the underlying spatial modes have significantly non-Gaussian histograms.

Under these assumptions, an easy calculation shows that the maximum likelihood estimate of the spatial unmixing matrix W_S is obtained by maximising the function

$$h_S(W_S) = \log |W_S| + \frac{1}{m} \sum_{ij} \log p_i(S_{ij}) \qquad (12.11)$$

We will need to know later that this quantity can be regarded as a measure of differential entropy [2].

With an appropriate choice of p_i, sICA can be shown to recover independent spatial modes if these exist. It is remarkable that the pdfs p_i need not be known exactly; sufficient conditions are given in [1]. In practice, it usually proves sufficient that the kurtosis of each p_i has the same sign as that of the true distribution. The 'vanilla' Bell and Sejnowski choice

$$p(x) \propto \tanh'(x) = \operatorname{sech}^2(x) \qquad (12.12)$$

used in [6] is appropriate for recovery of spatial modes having histograms with positive kurtosis.

12.4.1 Minimisation Routines

All the analyses in this chapter were performed in MatLab using an optimised Gauss - Newton procedure based on analytic gradients, and stopped when the gradient magnitude was smaller than numerical errors and the Hessian matrix was positive definite.

To avoid long derivations we have not specified the gradients for the entropy expressions to be maximised. Analytic formulae for gradients are available from the authors. For reasonably-sized data sets, direct minimisation of negentropy using standard routines will be found to be comparable in speed to (and, we believe, much more reliable than) gradient descent routines, natural or otherwise.

12.4.2 Application of sICA to OI Data

Figure 12.3 shows the results of applying vanilla sICA to our optical imaging data. The results are disappointing. The images are, on visual inspection, possibly a little less cluttered than those from PCA and the third component

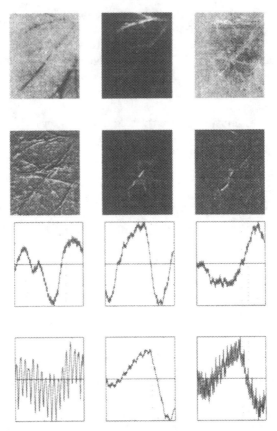

Fig. 12.3. Results of applying sICA to the optical imaging data set. The images in the top panel show the six spatial modes and the plots beneath their associated temporal variations.

shows weak evidence for a spatially localised response mode but none of the modes shows a clear temporal response to stimulation. (although the third mode seems to respond nicely about 1 s before stimulation starts!).

Notice that the motion artifact is still present as mode four but is now no longer associated with a clean periodic response. The spatial independence constraint alone is clearly not adequate to recover this mode accurately without placing some constraint on temporal behaviour.

The conclusion we draw from these results is that satisfactory physical modes have not been recovered because they are *inadequately modelled* by sICA assumptions. In particular, the constraint of spatial independence is both inadequate and inappropriate in this situation. In most practical problems it will be satisfied only approximately, if at all, and so it should be used in modelling as a minimum commitment hypothesis at best (for a lively but sometimes unreliable account of Bayesian methods and the role of minimum commitment hypotheses in estimation see [5]).

We propose to remedy these problems in two ways: first by specifying adequate prior information about temporal modes and secondly by changing our point of view on the interpretation of the statistical hypothesis underlying the sICA algorithm.

12.5   The Weak Causal Model

Since sICA places no prior constraints on temporal modes it neatly avoids the orthogonality problem of SVD but it is clearly not optimal in cases where prior information is available about temporal responses.

In model-driven analysis methods (such as generalised linear model analysis) this prior information must be given as an explicit temporal response curve or a low-order basis for such curves. Here we wish to consider situations in which the prior information does not completely specify the form of the temporal response. Such information can sometimes be available from experimental protocols.

For example, the response to a stimulus must follow stimulation. This causality constraint can be characterised in terms of the unpredictability of post-stimulus behaviour given pre-stimulus behaviour. The weak causal model *weak causal model* we describe below attempts first to quantify and then to optimise this unpredictability.

We assume that a low rank approximate factorisation $X = PQ^t$ is available. The first temporal mode in the parameterised factorisation equation 12.9 is a linear combination of columns of Q with coefficients from the first column of the temporal unmixing matrix

$$t = Qw \qquad w = \mathrm{col}_1(W_T) \tag{12.13}$$

Splitting the data set temporally into pre and post stimulus parts gives

$$t_{\mathrm{pre}} = Q_{\mathrm{pre}}w \qquad t_{\mathrm{post}} = Q_{\mathrm{post}}w \tag{12.14}$$

A simple model for predictability is to assume that, in the absence of stimulation, behaviour is constant up to additive noise. This is the only case we will consider here although the derivation below can easily be extended to other linear predictive models.

The constant pre-stimulus signal can be estimated from the mean of the pre-stimulus data

$$\hat{t} = \bar{t}_{\mathrm{pre}} = \overline{Q}_{\mathrm{pre}}w \tag{12.15}$$

(here overbars are used to denote means over time, hats to indicate estimators). The noise variance for this simple model can then be estimated from the pre-stimulus data as

$$E[(t_{\text{pre}} - \hat{t})^2] = w^t(Q_{\text{pre}} - \overline{Q}_{\text{pre}})^t(Q_{\text{pre}} - \overline{Q}_{\text{pre}})w = w^t C_2 w \qquad (12.16)$$

The magnitude of the post-stimulus response can now be characterised using the variance of the prediction error

$$E[(t_{\text{post}} - \hat{t})^2] = w^t(Q_{\text{post}} - \overline{Q}_{\text{pre}})^t(Q_{\text{post}} - \overline{Q}_{\text{pre}})w = w^t C_1 w \qquad (12.17)$$

and the signal-to-noise ratio

$$F(w) = \frac{w^t C_1 w}{w^t C_2 w} \qquad (12.18)$$

gives us a scale invariant measure of the size of the causal response. It can be regarded as an F-ratio statistic designed to test the hypothesis that the signal is stationary, the hypothesis being rejected for large F.

We can now recover the temporal mode which is maximally unpredictable by maximising $F(w)$ over all w. Since $F(w)$ has the form of a Rayleigh quotient, its maximum can be obtained by solving the generalised eigenvalue problem

$$C_1 w = \lambda C_2 w \qquad (12.19)$$

(this is the easy case where both matrices are symmetric and positive indefinite) and choosing the eigenvector w corresponding to the largest eigenvalue (see [3] for more details and related applications of the Rayleigh quotient).

Maximising $F(w)$ is equivalent to maximising the quantity

$$h_T(w) = \frac{1}{2} \log \frac{w^t C_1 w}{w^t C_2 w} \qquad (12.20)$$

which is an approximation (under Gaussian assumptions [4]) to the information capacity of the signal channel isolated by w. The fact that the weak model maximises a differential information measure will be important later.

12.5.1   Weak Causal Model Applied to the OI Data Set

The results of directly applying the weak causal model to obtain an optimal causal temporal response from the OI data set are not shown here but the response obtained is virtually identical to the first time series plot of figure 12.5 which shows a very satisfactory, biologically plausible, peaked response just after stimulation.

Much experimental and theoretical work is directed at verifying the detailed form of this response curve. For example, is there a hint of a causal component which shows a fast dip then a rise after stimulation? If such a dip component were present this might have useful implications for fMRI technology. Here we will restrict ourselves to a simpler question: is the causal response we have recovered significant, and if so, with what spatial region is it associated?

12.5.2   Some Remarks on Significance Testing

Since we have performed a multi-parameter optimisation to obtain the causal response curve it is not clear that the result is significant. In the absence of a properly specified Bayesian model, we follow, with some reluctance, orthodox statistical procedure, and ask how likely it is that a data set which is in fact stationary could produce such a striking result by chance.

In principle we could construct a parametric significance test based on the quantity F, however this approach would require Gaussian assumptions on the underlying signals and counting degrees of freedom might be complex. A more straightfoward approach would seem to be the use of a test based on randomisation of our data set.

One candidate surrogate data set, with which performance on the actual data can be compared, is obtained by simply randomising the initial data set in time and this will clearly destroy any causal structure in the data set. This is not a fair randomisation since most surrogate data examples would be highly discontinuous in time and it is easier to maximise the variance ratio F for signals correlated in time than for white signals (if we can arrange for the noise estimate $t_{\text{pre}} - \hat{t}$ to be small at a few points, its predictability ensures that it is also small close to those points).

If we restrict ourselves to testing first-order stationarity, we must construct surrogate data with the same auto-correlation stucture as the original. Such data can be obtained by phase randomisation of the original data set, treated as a vector random variable sampled over time. Phase randomised surrogate data sets have previously found applications in testing experimental data for the presence of chaotic behaviour.

Using phase randomised data sets in this way we found that the best causal component in the original data set had a causality F-ratio two orders of magnitude higher than that obtained from the best surrogate data set in 1000 trials. This would indicate that the causal response we recovered is significant at much better than the 1% level.

12.6   The Weak Periodic Model

If a stimulus has known period, for example in a box-car (ABAB...) experimental design, it is reasonable to look for a periodic response with the same period. A weak periodic model weak periodic model appropriate to this situation is easily derived.

For simplicity we assume that the period J is a whole number of time steps (the analysis can be extended to a non-integral period). The best periodic approximation to the time series $t = Qw$ (in the sense of least squares) has value at time j given by taking the mean of all the values at times $j \pm kJ$ for which we have observations. We denote this best periodic approximation by $\hat{t}$. Since the procedure is linear we have

$$\hat{t} = \hat{Q}w \tag{12.21}$$

(where the estimator above is applied to each column of Q).

The noise variance can then be estimated as

$$E[(t-\hat{t})^2] = w^t(Q-\hat{Q})^t(Q-\hat{Q})w = w^tC_2w \qquad (12.22)$$

and the signal variance as

$$E[t^2] = w^tQ^tQw = w^tC_1w \qquad (12.23)$$

so that the maximally periodic response can again be obtained by maximising a scale-invariant expression of the form.

$$h_T(w) = \frac{1}{2}\log F(w) = \frac{1}{2}\log\frac{w^tC_1w}{w^tC_2w} \qquad (12.24)$$

We have applied this technique succesfully to auditory fMRI data to localise response to stimulation in a box-car experimental design. It cannot be applied directly to our OI data set since the period J of the periodic component is unknown a priori (though candidate values could have been obtained by simultaneous monitoring of breathing or heart rate). In this case we can estimate the period by plotting the maximum of the statistic $F(w)$ as a function of the assumed period. If there is a component with period J, peaks will appear in this curve at its integer multiples $J, 2J, 3J, \ldots$.

The periodicity plot described above finds at least three significant periodic components in the optical imaging data set. We performed the analysis described here using only the strongest component (although recent work has shown that specifying all components can significantly improve the results of the regularised procedure described below).

The results of directly applying the weak periodic model to obtain an optimal periodic temporal response from the OI data set are not shown here but the response obtained is virtually identical to the second time series plot of figure 12.5 which shows a characteristic periodic response of rebound type.

12.7   Regularised Weak Models

The weak temporal models we have defined determine a unique time course but leave the associated spatial mode under constrained. To see this suppose

$$X = (s|\tilde{S})(t|\tilde{T})^t = st^t + \tilde{S}\tilde{T}^t \qquad (12.25)$$

is any factorisation including the optimal response t as one temporal component. Then

$$X = (s+\tilde{S}a)t^t + \tilde{S}(\tilde{T}-ta^t)^t = (s+\tilde{S}a|\tilde{S})(t|\tilde{T}-ta^t)^t \qquad (12.26)$$

is also a factorisation for any column $k - 1$ vector a. Thus the spatial mode s associated with t in the original factorisation can be mixed with *any* combination of other spatial modes. In particular we wish to emphasise that the correlation recipe

$$s = Xt^t \tag{12.27}$$

which is often used to find s, is only one of many possible choices, and will only be correct if t is uncorrelated with all other physical temporal modes, which is exactly the hypothesis we wish to avoid.

We propose that the independence measure used in spatial ICA be used to regularise this ill-posed problem. That is, we will attempt to recover the optimal temporal mode under the sICA assumptions on spatial modes by maximising a linear combination

$$h(W_S) = h_T(w) + \alpha h_S(W_S) \qquad w = \mathrm{col}_1(W_S^{-t}) \tag{12.28}$$

The coefficient α is potentially arbitrary and can be used to balance the effect of the temporal and spatial constraints. However we have noted that both the spatial and temporal components of this expression can be interpreted as channel entropies, and, since entropies are additive, it is tempting to hypothesise that the correct choice is $\alpha = 1$. In fact, this value works well in practice and has been used in all the examples.

We have introduced the sICA functional here as a regulariser for the ill-posed problem given by the temporal weak model. In fact, we originally developed the weak temporal model as a regulariser for sICA, which is itself ill-posed since the functional has a very shallow extremum in some dimensions for the data sets with which we are concerned.

12.8   Regularised Weak Causal Model Applied to OI Data

The results of applying a regularised sICA analysis to the optical imaging data set are shown in figure 12.4. The same spatial pdf is used for each mode (high kurtosis: $p \propto \tanh'$). Only two of the six modes recovered are shown in the figure.

The first mode clearly shows the optimally causal temporal response associated with a localised spatial structure in the barrel area. It is important to realise that though the first temporal response is constrained to be maximally causal by this procedure, the type of spatial activity map with which it is associated is not pre-determined by the choice of algorithm. Thus the analysis provides clear evidence for a localised causal response to stimulation. The second mode shows that the motion artifact is still recovered but with a distorted time course.

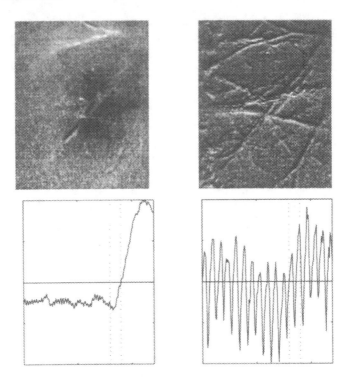

Fig. 12.4. Results for the weak causal model regularised by sICA applied to the OI data set. Only the causal response and motion artifact modes discussed in the body of the text are shown.

Further evidence for the choice $\alpha = 1$ in equation 12.28 is given by the fact that optimal causal mode is virtually unaffected by the inclusion of the sICA term, and so sICA is acting primarily to recover undetermined parameters, just as it should.

12.9   Image Goodness and Multiple Models

If truly independent components are not expected to be present then we prefer to regard the quantity h_S as a minimal commitment measure of image goodness. It measures how well the component image histograms fit the hypothesised pdf then independence is assumed only because we have no evidence for dependence. Under this interpretation, the choice of pdfs p_i can be expected to be critical.

The images we are interested in will usually contain localised features, such as barrel or vascular activity, on a homogeneous background. Such images are well characterised by the skewness (rather than the kurtosis) of their intensity histograms. We have found that a convenient functional form for the pdf is then

$$p(x; a, b) \propto \exp\left(\frac{a - b}{2}x - \frac{a + b}{2}\sqrt{x^2 + 1}\right) \qquad (12.29)$$

which has the asymptotic behaviours

$$p \approx e^{ax} \qquad x \to -\infty \qquad (12.30a)$$
$$p \approx e^{-bx} \qquad x \to +\infty \qquad (12.30b)$$

The constants a, b can be used to adjust the skewness of the distribution and the exponential decay rates mean that it shares the pleasant robustness properties of $\tanh'$.

A symmetric pdf such as $p \propto \tanh'$ is more appropriate for identifying components such as small-motion artifacts, which are essentially derivative images and usually contain both positive and negative values symmetrically. Of course it is possible to hypothesise different pdfs for different spatial modes. The choice of an appropriate p_i for each mode can then be regarded as a rudimentary statistical model of the underlying spatial structures.

If we wish to specify multiple temporal models we can add multiple weak model information terms into the equation to be maximised (12.28) . This procedure can produce inconsistent results when the recovered modes are correlated but is adequate for the application described below. A more powerful approach based on the information capacity of a multi-dimensional channel [4] seems to allow multiple correlated signals to be recovered effectively.

12.10   A Last Look at the OI Data Set

We combine all these techniques for a final look at the OI data set. We use the regularised ICA framework with spatial and temporal behaviour tuned to extract two modes of particular interest along with four unspecified modes.

The first mode is assumed to be a localised response to stimulation, this mode is modelled using a weak causal temporal model associated with a skew pdf for its spatial behaviour. The second mode is assumed to be a periodic pulse or breathing artifact associated with a derivative image, for which we use a weak periodic temporal model and a symmetric spatial pdf. The four other modes were allocated no temporal constraint; two were assumed to be spatially skew, two symmetric (nominal values $a = 4$, $b = 1$ were used in equation 12.29 to specify skewness).

Figure 12.5 shows the two temporally modelled modes recovered by this analysis. The use of the skew pdf for the causal response mode greatly increases the contrast between barrel and background in the localised spatial mode (though this may not be clearly evident in the reproduction) and a clean periodic response is now associated with the derivative image corresponding to a motion artifact.

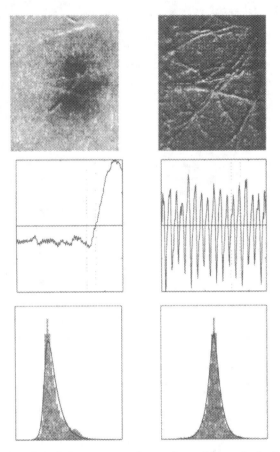

Fig. 12.5. Results of applying a composite weak model regularised by spatial ICA to the OI data set. Images in the top panel show the six spatial modes and the plots below their associated temporal variation with dotted lines bounding the period of whisker stimulation. Causal and periodic models respectively were assigned to the first two modes. The bottom panel of histograms show the fit between the image histograms and the assumed pdfs.

The bottom panel in Figure 12.5 shows the image histograms for these two spatial modes (shaded areas). These histograms agree well with the assumed symmetric and skew forms for the pdf (overlaid in black).

12.11   Conclusion

We would like to draw some general conclusions based on our experiences in applying ICA to real data sets:

- ICA is not a magic bullet. In fact, the nature of the hypotheses underlying ICA make its application to many real problems problematic.
- The independence assumption on which ICA is based may, to some extent, be a red herring. Rather than investigating the (admittedly elegant) properties of the ICA entropy measure when independence is present, it may be more profitable to look at its behaviour when the assumptions are violated. Our suggestion is that it still gives (as the older maximum entropy technique does) an optimal minimal commitment hypothesis based on a very simple statistical model of image generation.
- Models based on generic properties such as causality, periodicity and non-Gaussian spatial statistics (and on other generic properties not discussed here, such as continuity and predictability) can be a very powerful and flexible tool in spatio-temporal image analysis.
- It is probably fair to say that exploratory or data-driven techniques are widely distrusted in the clinical and experimental literature. This distrust is so strong that in some research areas it is difficult to publish results which are not shown to be significant using one of the standard model-based analysis packages. No analysis technique will be taken seriously in these areas until we can propose acceptable significance testing procedures.

References

1. S. Amari, T. Chen, A. Cichocki. *Neural Networks*, **10**: 1345-1351, 1995.
2. A. J. Bell and T. J. Sejnowski. An information-maximisation approach to blind separation and blind deconvolution. *Neural Computation* **7**(6): 1129-1159, 1995.
3. M. Borga. *Learning Multidimensional Signal Processing*. Dissertation, Linköping University, Sweden, 1998.
4. T. M. Cover and A. J. Thomas. *Information Theory*. Wiley Interscience, New York, 1991.
5. E. T. Jaynes. *Probability Theory: the Logic of Science* (Available from http://omega.albany.edu:8008/JaynesBook, 1995)
6. M. J. McKeown, S. Makeig, G. G. Brown, J. P. Jung, S. S. Kinderman and T. J. Sejnowski. *Proceedings of the National Academy of Sciences*, **95**: 803-811, 1998.
7. J. Mayhew, Y. Zheng, Y. Hou, B. Vuksanovic, J. Berwick, S. Asskew, P. Coffey. *Neuroimage*, **10**: 304-326, 1999.

13 Independent Components in Text

Thomas Kolenda, Lars Kai Hansen and Sigurdur Sigurdsson

13.1 Introduction

Automatic content-based classification of text documents is highly important for information filtering, searching, and hyperscripting. State-of-the-art text mining tools are based on statistical pattern recognition working from relatively basic document features such as term frequency histograms. Since term lists are high-dimensional and we typically have access to rather limited labeled databases, *representation* becomes an important issue. The problem of high dimensions has been approached with principal component analysis (PCA) - in text mining called latent semantic indexing (LSI) [4]. In this chapter we will argue that PCA should be replaced by the closely related independent component analysis (ICA). We will apply the ICA algorithm presented in Chapter 9 which is able to identify a generalizable low-dimensional basis set in the face of high-dimensional noisy data. The major benefit of using ICA is that the representation is better aligned with the content group structure than PCA. We apply our ICA technology to two public domain data sets: a subset of the MED medical abstracts database and the CRAN set of aerodynamics abstracts. In the first set we find that the unsupervised classification based on the ICA conforms well with the associated labels, while in the second set we find that the independent text components are stable but show less agreement with the given labels.

13.1.1 Vector Space Representations

The vector space model involves three steps: indexing, term weighting and a similarity measure [14]. Features are based on single word statistics; hence, first we create a term set of all words occurring in the database. This term set is screened for words that do not help to differentiate documents in terms of relevance. This is called *indexing*. In this *stop list* we find very frequent words like and, is and the. We also eliminate infrequent words that occur only in a few documents. The use of term frequency within a document to discriminate content-bearing words from the function words has been used for a long time [10]. Elimination of high-frequency terms is necessary as they can be very dominant in the term frequency histogram as shown in figure 13.1. When the term set for the document collection has been determined, each

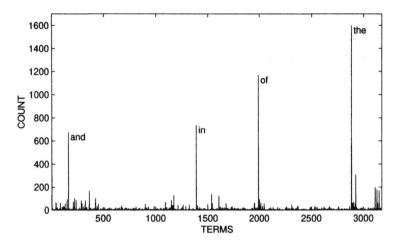

Fig. 13.1. Text mining is based on simple term frequency histograms. We show a histogram prior to screening for high frequency words. Note that common words like and, is and the totally dominate the histogram, typically without much bearing on the subsequent classification.

document can be described with a vector. For document j, the document vector is $D_j = [w_{1j}\ w_{2j}\ \cdots\ w_{tj}]^{\mathsf{T}}$, where t is the number of terms in the term list, i.e. the union of content-bearing words for all documents in the collection. We will form the term-document matrix for convenience, given by

$$
\underset{t \times d}{X} =
\begin{bmatrix}
w_{11} & w_{12} & \cdots & w_{1d} \\
w_{21} & w_{22} & \cdots & w_{2d} \\
\vdots & \vdots & & \vdots \\
w_{t1} & w_{t2} & \cdots & w_{td}
\end{bmatrix}
\tag{13.1}
$$

where d is the number of documents in the database.

Determining the values of the weights is called *term weighting*. There have been suggested a number of different term weighting strategies [15]. The weights can be determined from single documents independent of the other documents or by using database-wide statistical measures. The simplest term weighting scheme is to use the raw term frequency value as weights for the terms. If we assume that document length is not important for the classification, this vector can be normalized to unit length,

$$
w_{ij} = \frac{f_{ij}}{\sum_{j=1}^{d} f_{ij}}
\tag{13.2}
$$

where f_{ij} is the frequency of term i in document j.

The document *similarity measure* is usually based on the inner product of the document weight vectors, but other metrics can be argued for.

Figure 13.2 shows a normalized term by document matrix with function words removed. The data used for visualization are the first five groups in the MED data, which will be described later in this chapter. Only the ten terms with the highest occurrence variance are shown.

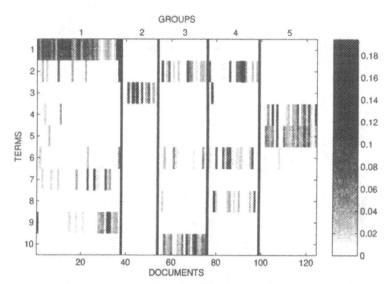

Fig. 13.2. The figure shows ten terms with the largest variances in the first five groups in the a document dataset. The columns are sorted by the group numbers from (1) to (5). Some of the terms are clearly "keywords".

13.1.2   Latent Semantic Indexing

All document classification methods that use single word statistics have well known language related ambiguities: polysemy and synonomy [4]. *Polysemy* refers to the problem of words that have more than one meaning. An example of this is the word jaguar which, depending on context, represents a sports-car or a cat. *Synonomy* is used to describe the fact that different words have similar meanings. An example of this are the words car and automobile.

Latent semantic indexing [4] is the PCA of the vector space model. The main objective is to uncover hidden linear relations between histograms, by rotating the vector space basis. If the major content differences form uncorre-lated (orthogonal) combinations, LSI will find them. The technique used for this transformation is the well known singular value decomposition (SVD). With the use of SVD, the term by document matrix X is decomposed into singular values and singular vectors, given by

$$\underset{t\times d}{X} = \underset{t\times r}{T} \cdot \underset{r\times r}{L} \cdot \underset{r\times d}{D^\top}$$

(13.3)

where r is the rank of X, L is a diagonal matrix of singular values and T and D hold the singular vectors for the terms and documents respectively. The terms and documents have been transformed to the same space with dimension r. The columns of T and D are orthogonal, i.e., uncorrelated. If X is full rank the dimension of the new space is d.

If the database is indeed composed from a few independent contents each characterized by a class histogram, we would expect relatively few relevant singular values, the remaining singular values being small and their directions representing noise. By omitting these directions in the term vector space we can improve the signal to noise ratio and effectively reduce the dimensionality of the representation. If the singular values are ordered by decreasing value, the reduced model using the k largest singular values is

$$\underset{t\times d}{X} \approx \underset{t\times d}{\widehat{X}} = \underset{t\times k}{T} \cdot \underset{k\times k}{L} \cdot \underset{k\times d}{D^\top}$$

(13.4)

where $k < r$, and $\widehat{X}$ is the rank k estimate of X. The selection of the number of dimensions, or k, is not trivial. The value of k should be large enough to hold the latent semantic structure of the database, but at the same time we want it as small as possible to obtain the optimal signal to noise ratio.

In Figure 13.3, the ten largest principal components of the DL matrix are shown using the MED data. In the upper row of Figure 13.4 we show scatterplots of projections on PC2 vs. PC1 and PC2 vs. PC3, and note that the documents (color coded according to class) fall in non-orthogonal rays emanating from origo. This strongly suggests the use of ICA. Decomposing the same data along a non-orthogonal basis as identified by ICA is shown in the middle row of Figure 13.4.

The use of ICA for information retrieval was earlier proposed by Isbell and Viola [7]. This approach focuses on the identification of sparse representation. In this paper we are not going to discuss the sparse nature of the ICA basis; we are mainly interested in the independence property and the ability of the independent components to perform unsupervised identification of relevant content class structures.

13.2   Independent Component Analysis

Reconstruction of statistically independent components or sources from linear mixtures is relevant to many information processing contexts, see e.g. [9] for an introduction and a recent review. The term by document matrix is considered a linear mixture of a set of independent sources (contents) each activating its characteristic semantic network. The semantic networks take

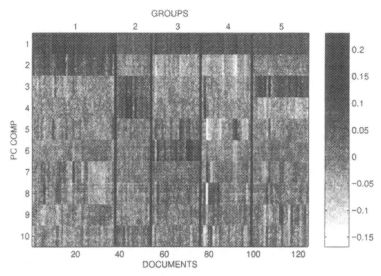

Fig. 13.3. The figure shows the ten first principal components of the $D \cdot L$ matrix for the first five groups in the MED dataset. The columns are sorted by the groups (1) to (5). The first components are clearly assigned to specific groups in the data set.

the form of non-orthogonal term occurrence histograms. Translating to the more conventional use of ICA for speech separation, terms play the role of "microphones" and the document index corresponds to the time index, while the independent content components correspond to "speakers".

13.2.1   Noisy Separation of Linear Mixtures

We review an approach to the source separation based on the likelihood formulation, see e.g., [2,12,11]. The specific model investigated here is a special case of the general framework proposed by Belouchrani and Cardoso [2], however we formulate the parameter estimation problem in terms of the Boltzmann learning rule, which allows for a particular transparent derivation of the mixing matrix estimate. We focus on generalisability of the ICA representation, and use the generalisation error as a means for optimizing the complexity of the representation viz. the number of sources.

Let the term by document matrix be the observed mixture signals denoted X, a matrix of size $t \times d$, where t is the number of terms in the word histogram and d is the number of documents. The noisy mixing model takes the form,

$$X = AS + U \tag{13.5}$$

where S is the source signal matrix (size $c \times d$, c is the number of sources), A is the $t \times c$ mixing matrix, while U is a matrix of noise signals with a param-

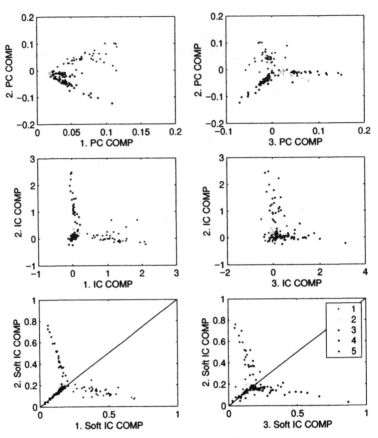

Fig. 13.4. Analysis of the MED set of medical abstracts, labeled in five classes here coded in colors. The two upper panels show scatterplots of documents in the latent semantic or principal component basis. In the middle panels we show the document location as seen through the ICA representation. Note that while the group structure is clearly visible in the PCA plots, only in the ICA plots is the group structure aligned with independent components. In the lower panels is the result of passing the IC components through softmax for classification. The diagonal is a simple classification decision boundary.

eterised distribution. The properties of the source signals are introduced by a parameterised prior distribution $P(S|\psi)$. The likelihood of the parameters of the noise distribution, the parameters of the source distribution and of the mixing matrix is given by,

$$L(A, \theta, \psi) = P(X|A, \theta, \psi) = \int P(X - AS|\theta)P(S|\psi)dS \qquad (13.6)$$

where $P(.|\theta)$ is the parameterised noise distribution. We assumed i.i.d. sources in equation 13.6 and we will assume also that the noise can modeled by i.i.d. Gaussian variables with variance $\theta = \sigma^2$,

$$P(U|\sigma^2) = \frac{1}{(2\pi\sigma^2)^{td/2}} \exp\left(-\frac{1}{2\sigma^2}\sum_{\tilde{t},\tilde{d}} U_{\tilde{t},\tilde{d}}^2\right) \tag{13.7}$$

where $\tilde{t}, \tilde{d}$ index terms and document variables. Finally, we will assume the parameter free source distribution of [11],

$$P(S) = \frac{1}{\pi^{dc}} \exp\left(-\sum_{\tilde{c},\tilde{d}} \log\cosh S_{\tilde{c},\tilde{d}}\right) \tag{13.8}$$

The sum runs over all sources ($\tilde{c} = 1, ..., c$) and documents ($\tilde{d} = 1, ..., d$).

Let us first address the problem of estimating the sources if the mixing parameters are known, i.e., for given A, σ^2. We use Bayes formula $P(S|X) \propto P(X|S)P(S)$ to obtain a posterior distribution of the sources

$$P(S|X, A, \sigma^2) \propto \exp\left(-\frac{1}{2\sigma^2}\sum_{\tilde{t},\tilde{d}}(X - AS)_{\tilde{t},\tilde{d}}^2 - \sum_{\tilde{c},\tilde{d}} \log\cosh S\right) \tag{13.9}$$

The *maximum a posteriori* (MAP) source estimate is found by maximising this expression w.r.t. S, providing the following non-linear equation to solve iteratively for the MAP estimate $\widehat{S}$,

$$-A^{\top}A\widehat{S} + A^{\top}X - \sigma^2\tanh\widehat{S} = 0 \tag{13.10}$$

There are two problems in equation 13.10. First, the equation is non-linear – though only weakly non-linear for low noise levels. Secondly, the "system matrix", $A^{\top}A$, may be ill-conditioned or even singular. A useful rewriting that takes care of potential ill-conditioning of the system matrix is given by,

$$\widehat{S} = \left(A^{\top}A + \sigma^2\right)^{-1}\left(A^{\top}X + \sigma^2\left(\widehat{S} - \tanh(\widehat{S})\right)\right) \tag{13.11}$$

This form suggests an approximate solution for low noise levels

$$\widehat{S} = S^0 + \sigma^2 H^{-1}\left(S^0 - \tanh S^0\right)$$
$$S^0 = H^{-1}A^{\top}X, \quad H = A^{\top}A + \sigma^2 \tag{13.12}$$

exposing the fact that the presence of additive noise turns the otherwise linear separation problem into a non-linear one.

Since the likelihood is of the hidden-Gibbs form we can use a generalised Boltzmann learning rule to find the gradients of the likelihood of the parameters A, σ^2. These averages can be estimated in a mean field approximation

[13,6] leading to recursive rules for A and σ^2,

$$\widehat{A} = X\widehat{S}^\top \left(\widehat{S}\widehat{S}^\top + \beta \mathbf{1}\right)^{-1} \tag{13.13}$$

$$\widehat{\sigma^2} = \frac{1}{td}\text{Tr}_t(X - \widehat{A}\widehat{S})^\top(X - \widehat{A}\widehat{S}) \tag{13.14}$$

β is a lumped effect of fluctuations neglected in the mean field approach. Fluctuation corrections (hence the magnitude of β) can be derived in the low noise limit, based on a Gaussian approximation to the likelihood [6].

13.2.2   Learning ICA Text Representations on the LSI Space

As we typically operate with 1000+ words in the terms list and fewer documents, we face a so-called *extremely ill-posed learning problem* [8] which can be "cured" without loss of generality by PCA projection. PCA decomposes the term by document matrix on eigen-histograms. These eigen-histograms are subject to an orthogonality constraint being eigenvectors to a symmetric real matrix. As noted above, we are interested in a slightly more general separation of sources that are independent as sequences, but not necessarily orthogonal in the word histogram, i.e., we would like to be able to perform a more general decomposition of the data matrix, corresponding to the model in equation 13.5. Before performing the ICA, we can make use of the PCA for simplification of the ICA problem. We first note that the likelihood, considered a function of the columns of A (histograms) can be split into two parts: part A_1 is orthogonal to the subspace spanned by the d rows of X and part A_2 is situated in the subspace spanned by the d columns of X. The first part is trivially minimised for any non-zero configuration of sources by putting $A_1 = 0$. It simply does not "couple" to data. The remaining part A_2 can be projected onto a d-dimensional hyperplane spanned by the documents. In this way we reduce the high-dimensional separation problem to the separation of a square (projected) data matrix of size $d \times d$. We note that it may often be possible to further limit the dimensionality of the PCA subspace, hence further reducing the histogram dimensionality t of the remaining problem.

As described earlier, the LSI model is merely performing a PCA on top of the vector space model and thus learning the ICA text representation can be viewed as a post-processing step for the LSI model. Inserting the ICA decomposition into equation 13.4 we decompose the term by document matrix into,

$$\underset{t\times d}{X} = \underset{t\times d}{T} \cdot \underset{d\times c}{A} \cdot \underset{c\times d}{S} \tag{13.15}$$

where T holds the term eigenvectors, A is the IC document projections on the PC basis and S holds the separated "contents".

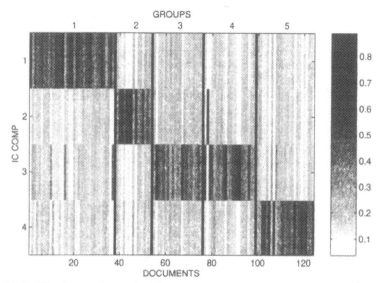

Fig. 13.5. The figure shows the IC components with four channels using the first five groups in the MED dataset. The columns are sorted by the group numbers from (1) to (5). The channel value is clearly related to the class number.

An example of the S matrix is shown in figure 13.5 using the MED data. The number of independent components is four, which will be shown to be the generalisation optimal. The S matrix is normalized with softmax so the outputs can be interpreted as the probability of a document belonging to each class. In this case, the unsupervised ICA is able to determine a group structure which very closely coincides with the "human" labels 1, 2, 5 but lumps groups 3 and 4 into one. Interestingly, running ICA with five or more component does not resolve groups 3 and 4 but rather finds a independent mixtures of the two.

13.2.3   Document Classification Based on Independent Components

To quantify the ability of ICA to group documents we convert the separated signal to "class probabilities" using the standard *softmax* normalization on the recovered source signals,

$$\phi_{\tilde{c}d} = \frac{\exp(S_{\tilde{c}d})}{\sum_{1=\tilde{c}}^{c} \exp(S_{\tilde{c}d})} \tag{13.16}$$

This is analogous to similarity based on projections onto principal components as used in LSI [4]. Since the ICA classification is unsupervised we need to match classes when comparing with manual labels.

13.2.4   Keywords from Context Vectors

Finding characteristic *keywords* to describe the context in a given independent component can be obtained by back projection of the documents to the original vector histogram space. This amounts to projection on to the identity matrix through the PCA and ICA bases. From equation 13.15 we find that TA is the basis change where columns represent the weight of the terms in each output (see Figure 13.6).

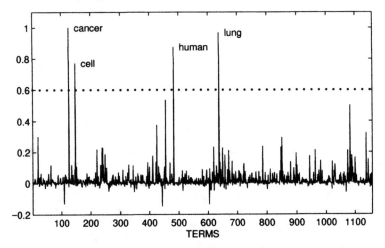

Fig. 13.6. In analysis of the MED data set, keywords can be found by back projection for a given component. Keywords above a specified threshold are chosen as words that best describe the given component's context.

Depending on how many and their weight, we choose the keywords above a specified threshold after normalizing, as in table 13.1.

13.2.5   Generalisation and the Bias-Variance Dilemma

The parameters of our ICA model are estimated from a finite random sample, and therefore they too are random variables inheriting noise from the dataset on which they were trained. Within the likelihood formulation the generalization error of a specific set of parameters is given by the average

Table 13.1. Confusion matrix and keywords from classification of MED with 2 to 5 output IC components. The confusion matrix compares the classification of the ICA algorithm to the labeled documents. Each IC component likewise produced a set of keywords, that are ordered by the size of the projection starting with the largest.

	C_1	C_2	C_3	C_4	C_5	keywords
IC_1	36	1	0	0	2	lens crystallin
IC_2	1	15	22	23	24	cell lung tissue alveolar normal cancer human

	C_1	C_2	C_3	C_4	C_5	keywords
IC_1	36	0	0	0	0	lens crystallin
IC_2	0	16	0	1	24	fatty acid blood glucose oxygen free maternal plasma level tension newborn
IC_3	1	0	22	22	2	cell lung tissue alveolar normal

	C_1	C_2	C_3	C_4	C_5	keywords
IC_1	36	0	0	0	0	lens crystallin
IC_2	0	16	0	1	0	oxygen tension blood cerebral pressure arterial
IC_3	1	0	22	21	2	cell lung tissue alveolar normal
IC_4	0	0	0	1	24	fatty acid glucose blood free maternal plasma newborn fat level

	C_1	C_2	C_3	C_4	C_5	keywords
IC_1	35	0	0	0	0	lens crystallin
IC_2	0	16	0	1	0	oxygen tension blood cerebral pressure arterial
IC_3	2	0	15	10	0	cells alveolar normal
IC_4	0	0	7	12	2	cancer lung human cell growth tissue found virus acid previous
IC_5	0	0	0	0	24	fatty acid glucose blood free maternal plasma newborn level fat

negative log-likelihood,

$$\Gamma(A, \theta, \psi) = \int P_*(X)[-\log \int P(X - AS|\theta)P(S|\psi)dS]dX \qquad (13.17)$$

$P_*(X)$ is the true distribution of data. The generalisation error is a principled tool for model selection. In the context of blind separation, the optimal number of sources retained in the model is of crucial interest. We face a typical

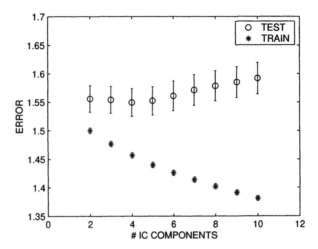

Fig. 13.7. ICA analysis of the MED set of medical abstracts. Training and test errors as functions of the dimensionality of the mixing matrix. The generalisation error shows a shallow minimum for four independent components, reflecting a bias-variance tradeoff as function of the complexity of the estimated mixing matrix.

bias-variance dilemma [5]. If too few components are used, a structured part of the signal will be lumped with the noise, hence leading to a high generalisation error because of "lack of fit". On the other hand, if too many sources are used we expect "overfit" since the model will use the additional degrees of freedom to fit non-generic details in the training data. The generalisation error in equation 13.17 can be estimated using a test set of data independent of the training set.

13.3   Examples

We will illustrate the use of ICA in text mining on two public domain data sets both available on the WWW [16]. The MED dataset has been known to produce good results in most search and classification models and therefore serves as a good starting point. The second dataset CRAN is a more complex set of documents with overlapping class labels and less obvious group structure.

   In general when constructing the histogram term by document matrix, a word that occurred in more than one document and was not present in a given list of stop words, was chosen as a term word. The length of each document histogram (the columns) was normalized to one to remove the effect of the document length. Using the "cure for extremely ill-posed learning" method based on PCA, the data matrix was reduced to a $d \times d$ problem without loss of generality.

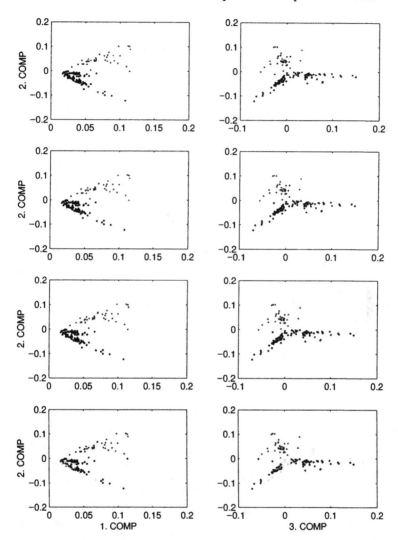

Fig. 13.8. The MED data set of medical abstracts. The data set consists of 124 documents in five topics. The "source signals" recovered in the ICA have been converted by a simple softmax classifier, and we have coded these classes by different colors. From top to bottom we show scatterplots in the principal component representation PC2 vs PC1 and PC2 vs PC3, with colors indicating the classification proposed by the ICA with 2,3,4,5 independent components respectively.

When converting ICA recovered sources to classifications using equation 13.16 we also matched the unsupervised ICA classes to the manual labels.

13.3.1   MED Data Set

The MED data set is a commonly studied collection of medical abstracts. In all, it consists of 1033 abstracts, of which 30 labels have been applied to 696 documents. For simplicity, we here consider 124 abstracts associated with the first five groups in the MED data set. 1159 terms were used to build the term by document matrix.

Here is a brief outline of the abstract groups:

1. The crystalline lens in vertebrates, including humans.
2. The relationship of blood and cerebrospinal fluid oxygen concentrations or partial pressures. A method of interest is polarography.
3. Electron microscopy of lung or bronchi.
4. Tissue culture of lung or bronchial neoplasms.
5. The crossing of fatty acids through the placental barrier. Normal fatty acid levels in placenta and fetus.

Results. In Figure 13.7 we show the test and training set errors evaluated on training sets of 104 patterns randomly chosen among the set of 124. The test set consists of the remaining 20 documents in each resample (ten-fold crossvalidation). The generalisation error shows a shallow minimum for four independent components, reflecting a bias-variance tradeoff as a function of the complexity of the estimated mixing matrix.

In Figure 13.4 we show scatterplots in the largest principal components and the most variant independent components. While the distribution of documents forms a rather well-defined group structure in the PCA scatterplots, clearly the ICA scatterplots are much better aligned to the axes. We conclude that the non-orthogonal basis found by ICA better "explains" the group structure. To further illustrate this finding, we have converted the ICA solution to a pattern recognition device by a simple heuristic. Normalizing the IC output values through softmax showed evidence that comparing the magnitude of the recovered source signals produced a method for unsupervised classification.

In Table 13.1 we show that this device is quite successful in recognizing the group structure although the ICA training procedure is completely unsupervised. For an ICA with three independent components two are recognized perfectly, and three classes are lumped together. The four component ICA, which is the generalisation optimal model, recognizes three of the five classes almost perfectly and confuses classes 3 and 4. Inspecting the groups we found that the two classes are indeed on very similar topics (they both concern medical documents on diseases of the human lungs) and investigating classifications for five or more ICA components did not resolve the ambiguity between them. The ability of the ICA-classifier to identify the topic structure is further illustrated in Figure 13.8 where we show the PC scatterplots color coded according to ICA classifications.

Finally, we inspect the components produced by ICA by back-projection using the PCA basis. Thresholding the ICA histograms, we find the salient terms for the given component. These terms are keywords for the given topic as shown in Table 13.1.

13.3.2   CRAN Data Set

The CRAN data set is a collection of aerodynamic abstracts. In all, it consists of 1398 abstracts with 225 different labels and some were labeled as belonging to more than one group. Furthermore, inspecting the abstracts we found a greater content class overlap, hence we expect discrimination to be much harder. Because of the high number of classes, some clusters were very small and we selected five content classes with 138 documents. In those groups the overlap was especially present in class 1 with 3 and class 2 with 5. A total of 1115 terms were used in the term by document matrix.

Here is a brief description of the abstracts groups:

1. The structural and aeroelastic problems associated with flight of high speed aircraft.
2. Calculating the effect and magnitude of the boundary-layer on wing pressure.
3. The similarity laws that must be obeyed when constructing aeroelastic models of heated high speed aircraft.
4. Calculating the aerodynamic performance of channel flow ground effect machines.
5. Summarizing theoretical and experimental work on the behaviour of a typical aircraft structure in a noise environment; is it possible to develop a design procedure?

Results. To recognize possible group structure we show scatterplots for the first three PC and IC components in Figure 13.9. Classes that overlap are marked with both a dot and a circle having colors representing their classes. Comparing the CRAN PC scatterplots with the MED scatterplots in Figure 13.4 it is clear that the CRAN data is much more heterogeneous in content. From Figure 13.9 it is clear that ICA has identified some group structure but not as convincingly as in the MED data. This is also borne out in Figures 13.12 and 13.13 image the document projection relations. The CRAN data is split into training and test sets: 100 abstracts are used for training and 38 for testing. The generalisation results in Figure 13.10 are obtained using 10-fold cross-validation. The test error shows a shallow minimum at three independent components, reflecting the bias-variance trade-off as a function

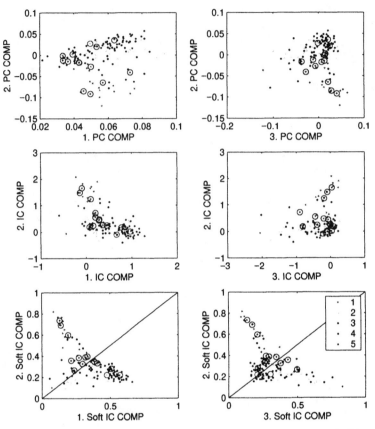

Fig. 13.9. Analysis of the CRAN set labeled in five classes here coded in colors. The two upper panels show scatterplots of documents in the Latent Semantic or Principal Component basis. In the middle panels we show the document location as seen through the ICA representation. Note that while the group structure is clearly visible in the PCA plots, only in the ICA plots is the group structure aligned with independent components. In the lower panels is the result of putting the IC components through softmax for classification. The diagonal line shows the decision boundary.

of the complexity of the estimated mixing matrix. The classification confusion matrix is found in Table 13.2. If we focus on the three source solution (the best from a generalization point of view) we find that ICA isolates class 1 and IC1, while the expected overlap between classes 2 and 5 is seen in IC2. Class 4 is placed in a component overlapping with classes 2 and 3 in IC3.

In Table 13.3 we have illustrated the classification consistency among ICAs with different numbers of components. First we adapted a five component system and we recorded the ICA class labels for each document. We

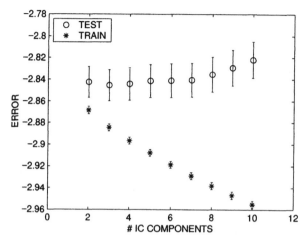

Fig. 13.10. ICA analysis of the CRAN set of aerodynamic abstracts. Training and test errors as function of the dimensionality of the mixing matrix. The generalisation error shows a shallow minimum for three independent components, reflecting a bias-variance tradeoff as function of the complexity of the estimated mixing matrix.

next adapted ICA's with two, three, and four components and created class labels. The confusion matrices show that although the ICA "unsupervised" labels are only in partial agreement with the manual labels they are indeed consistent and they show a hierarchical structure similar to the MED data.

13.4   Conclusion

We have shown how independent component analysis can be used to identify a compact and content relevant basis in the term frequency histogram space. We used an ICA framework based on maximum likelihood estimation which allows us to discuss generalizability of the estimated models. This provides the possibility of unsupervised estimation of the number of independent sources or the present context of the number of independent content classes. On two small labeled abstract databases we found that unsupervised classification using the recovered independent ICA sources compared well with the given set of labels. In the so-called MED data, three out of four classes complied perfectly while two remaining classes were lumped together. A closer examination showed that the lumped classes indeed concerned closely related subjects. The CRAN database is more complicated in that the abstracts concern very closely related subjects, and that the associated manual labels are overlapping (abstracts belong to several classes). The ICA representa-

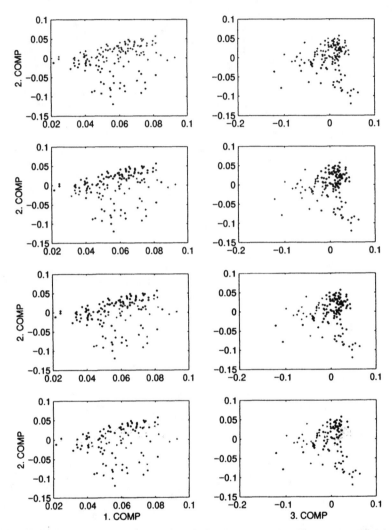

Fig. 13.11. The CRAN dataset of aerodynamic abstracts. The dataset consists of 138 documents in five topics. The "source signals" recovered in the ICA has been converted to a simple classifier, and we have coded these classes by different colors. From top to bottom we show scatterplots in the principal component representation 1 vs. 2 and 3 vs. 2., with colors signifying the classification proposed by the ICA with 2,3,4,5 independent components respectively.

tion found a consistent group structure which only partially agreed with the manual labeling.

Table 13.2. Confusion matrix and keywords from classification of CRAN with 2 to 5 output IC components. The confusion matrix compares the classification of the ICA algorithm to the labeled documents. Each IC component likewise produced a set of keywords, that are ordered by the size of the projection starting with the largest.

	C_1	C_2	C_3	C_4	C_5	keywords
IC_1	23	0	12	7	0	flutter panel
IC_2	2	25	6	26	37	flow body pressure mach theory

	C_1	C_2	C_3	C_4	C_5	keywords
IC_1	19	0	4	3	0	flutter panel
IC_2	2	16	6	16	37	flow pressure body mach number shock hypersonic
IC_3	4	9	8	14	0	wing body

	C_1	C_2	C_3	C_4	C_5	keywords
IC_1	17	0	3	3	0	flutter panel
IC_2	2	13	6	11	34	flow pressure mach number hypersonic shock heat layer body boundary transfer
IC_3	5	0	9	0	1	wing thermal temperature stress aerodynamic supersonic
IC_4	1	12	0	19	3	wing body theory lift flow

	C_1	C_2	C_3	C_4	C_5	keywords
IC_1	17	0	3	3	0	flutter panel
IC_2	1	11	0	14	1	wing body lift theory
IC_3	5	0	11	0	0	thermal wing temperature stress heat
IC_4	0	7	1	12	24	flow body
IC_5	2	7	3	4	12	mach pressure number heat

The ICA representation is based on independence of the recovered sources; hence, independence of contents as seen through word frequency histograms. This may or may not be well aligned with a manual labeling. If the ICA group structure doesn't comply with manual labeling it informs us about a potentially useful hidden structure in the text data.

In future work we will examine the viability of the ICA basis for more complex supervised adaptive classifiers. We expect that the approximate independence of the ICA representation will assist classifier learning.

Table 13.3. Confusion matrix from classification of CRAN with 2 to 5 output IC components. The confusion matrix compares the classification of the ICA algorithm to the five ICA estimated classes.

	C_1	C_2	C_3	C_4	C_5	keywords
IC_1	23	5	13	0	1	flutter panel
IC_2	0	22	3	44	27	flow body pressure mach theory number

	C_1	C_2	C_3	C_4	C_5	keywords
IC_1	23	0	2	0	1	flutter panel
IC_2	0	21	11	3	0	wing body
IC_3	0	6	3	41	27	flow pressure body mach number shock hypersonic

	C_1	C_2	C_3	C_4	C_5	keywords
IC_1	23	0	0	0	0	flutter panel
IC_2	0	26	0	7	1	wing body theory lift flow
IC_3	0	0	13	0	2	wing thermal temperature stresses
IC_4	0	1	3	37	25	flow pressure mach number hypersonic shock heat layer body boundary transfer

	C_1	C_2	C_3	C_4	C_5	keywords
IC_1	23	0	0	0	0	flutter panel
IC_2	0	27	0	0	0	wing body lift theory
IC_3	0	0	16	0	0	thermal wing temperature stress heat
IC_4	0	0	0	44	0	flow body
IC_5	0	0	0	0	28	mach pressure number heat

Acknowledgment

This work was funded by the Danish Research Councils through the Intermedia plan for multimedia research and the THOR Center for Neuroinformatics.

References

1. B. T. Bartell, G. W. Cottrell, R. K. Belew. Latent Semantic Indexing is an optimal special case of multidimensional scaling, *SIGIR Forum, ACM Special Interest Group on Information Retrieval*, 161 - 167, 1992.
2. A. Belouchrani and J.-F. Cardoso. Maximum likelihood source separation by the expectation-maximization technique: deterministic and stochastic implementation. *In Proc. NOLTA*, 49-53, 1995.

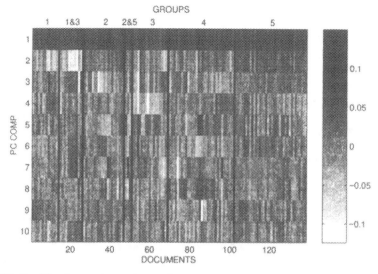

Fig. 13.12. The figure shows the 10 first principal components of the $D \cdot L$ matrix for the groups (1) to (5) and (1&3) and (2&5) in the CRAN data set. The columns are sorted by groups. Relations between principal components and groups can be observed, e.g., the second principal component seems to represent group (1) and (1&3).

3. J-F. Cardoso. Infomax and maximum likelihood for blind separation. *IEEE Sig. Proc. Letters*, **4**(4):112–114, 1997.

4. S, Deerwester, S. T. Dumais, G. W. Furnas, T. K. Landauer, R. Harshman. Indexing by Latent Semantic Analysis. *J. Amer. Soc. for Inf. Science* **41**, 391-407, 1990.

5. S. Geman, E. Bienenstock, and R. Doursat. Neural Networks and the Bias/Variance Dilemma, *Neural Computation*, **4**: 1-58, 1992.

6. Hansen L.K. Blind separation of noisy mixtures. Department of Mathematical Modeling, Tech. Univ. Denmark, http://eivind.imm.dtu.dk/pub, 1998.

7. C. L. Isbell and P. Viola. Restructuring sparse high dimensional data for effective retrieval. *Advances in Neural Information Processing Systems* **11**, 480 - 486, 1998.

8. B. Lautrup,L. K. Hansen, I. Law, N. Mørch, C. Svarer, S. C. Strother. Massive weight sharing: A cure for extremely ill-posed problems. In H.J. Hermanet al., editors. *Supercomputing in Brain Research: From Tomography to Neural Networks.* World Scientific Pub. Corp., 137 - 148, 1995.

9. T.-W. Lee. Independent component analysis: theory and applications, *Kluwer Academic Publishers*, 1998.

10. H. P. Luhn. The automatic creation of literature abstracts. *IBM Journal of Research and Development* **2**(2): 159 - 165, 1958.

11. D. MacKay. Maximum likelihood and covariant algorithms for independent components analysis. "Draft 3.7", 1996.

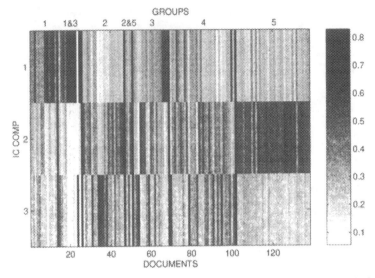

Fig. 13.13. The figure shows the IC components after softmax using 4 channels in the CRAN data set. The columns are sorted by group. Groups (1), (1&3) and (5) clearly visible in the channels, but the other groups overlap.

12. B. A. Pearlmutter and L. C. Parra. A context-sensitive generalization of ICA. *Proc. International Conference on Neural Information Processing*, 1996.
13. C. Peterson & J.R. Anderson. A mean field theory learning algorithm for neural networks. *Complex Systems*. **1**: 995 - 1019, 1987.
14. G. Salton. Automatic text processing: The transformation, analysis, and retrieval of information by computer, *Addison-Wesley*, 1989.
15. G. Salton and C. Buckley. Term weighting approaches in automatic text retrieval, *Department of Computer Science, Cornell University, Technical Report TR87-881*, 1987.
16. Smart (1999): ftp.cs.cornell.edu/pub/smart/, Department of Computer Science, Cornell University, public ftp

14 Seeking Independence Using Biologically-Inspired ANN's

Pei Ling Lai, Darryl Charles and Colin Fyfe

14.1 Introduction

In this chapter, we make a case for research into the use of biologically-inspired artificial neural networks (BIANNs) in the search for independence. This may seem an odd case to have to make in a book derived from a workshop which followed the Seventh International Conference on Artificial Neural Networks (ICANN99), the most prestigious ANN conference in Europe. However, as this book demonstrates, many of the methods currently being investigated by the neural network community are very different from the biologically-inspired networks which we will advocate.

The case for BIANNs is simple but strong: we still do not have ICA machines which are even remotely as good at extracting single sources from mixtures as humans are. Let us recall the lessons from early research into artificial intelligence: early AI was based on symbol processing and has generated great successes - artificial chess players, theorem provers, and expert systems are all impressive achievements - but these successes are as nothing compared to the intelligence of a typical three-year-old child. To create robust creative intelligences, we must emulate nature. This, of course, was one of the major forces behind much early research into ICA [16] in which is described a 'neuro-mimetic architecture'.

The case against this of course states that we do not know which features of human brains are crucial in creating intelligence: humans can deal with information which is only probably true and so Bayesian methods can be used on such problems; humans learn very quickly and so fast batch methods may be used on these problems and so on. Our argument against this is again simple: until we know what features of human brains are essential, we should stick closely to the known features of biological neural networks.

The aspect of BIANNs on which we focus in this chapter is that of learning - our networks will use simple Hebbian and anti-Hebbian learning to find interesting statistical features in our data sets. In particular, we shall review three networks which search for independence in data sets and in doing so, we shall discuss problems other than the standard ICA problem.

The remainder of this chapter is structured as follows: in section 14.2, we review a negative feedback neural network which has been shown to perform principal component analysis. In section 14.3, we extend this network to a factor analysis network and show that it is capable of identifying independent causes in a data set. In section 14.4, we discuss a canonical correlation network and show how its nonlinear version can be used to identify dependencies between two data sets. In Section 14.5, we optimise the basic network of Section 14.2 for non-Gaussian noise and show how it may be used for extraction of independent sources. We conclude with a network which performs robust minor component analysis and show its ability to de-noise when Signal to Noise Ratios are very small.

14.2 The Negative Feedback Network

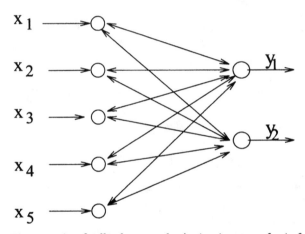

Fig. 14.1. The negative feedback network. Activation transfer is fed forward and summed and returned as inhibition. Adaption is performed by simple Hebbian learning.

The statistical properties elicited from the learning methods of many BIANNs are well known (for a review see [24]). In particular, those correlation finding networks which are adapted using Hebbian learning have attracted widespread interest; such networks are used for finding the principal components of the input data. Principal component analysis (PCA) is a statistical technique which finds the best linear compression of a data set and many artificial neural networks (e.g. [21,22,25,23]) have been developed to perform this type of data compression.

Figure 14.1 shows a network which can be used for PCA: the data is fed forward from the input neurons (the x-values) to the output neurons. Here the weighted summation of the activations is performed and this is fed back via the same weights and used in a simple Hebbian learning procedure. Consider a network with N dimensional input data and having M output neurons. Then the output of the ith output neuron is given by

$$y_i = \text{act}_i = \sum_{j=1}^{N} w_{ij} x_j \tag{14.1}$$

where x_j is the activation of the jth input neuron, w_{ij} is the weight between this and the ith output neuron and act_i is the activation of the ith neuron. This firing is fed back through the same weights as inhibition to give

$$x_j(t+1) \leftarrow x_j(t) - \sum_{k=1}^{M} w_{kj} y_k \tag{14.2}$$

where we have used (t) and $(t+1)$ to differentiate between activation at times t and $t+1$. Now simple Hebbian learning between input and output neurons gives

$$\Delta w_{ij} = \eta_t y_i x_j(t+1)$$
$$= \eta_t y_i \{ x_j(t) - \sum_{l=1}^{M} w_{lj} y_l \}$$

where η_t is the learning rate at time t. This network has previously been used in [9] to perform PCA. This network actually only finds the subspace spanned by the principal components; we can find the actual Principal Components by introducing some asymmetry into the network [9].

The network as described above seems non-BIANN in that the same weights are being used for forward and backward activation transfer, however it has been shown [8] that we may introduce a second set of weights which will handle the feedback and that both sets of weights co-learn the principal component directions. Intuitively, since both sets of weights use simple Hebbian learning and have access to the same information, both will learn the same thing.

However PCA only allows us to decorrelate the outputs and we are interested in this book in creating independent outputs: the above network uses only second order statistics and so we may only decorrelate the outputs whereas we wish to use higher order statistics (see below) to find independent sources.

We have therefore extended the network by introducing non-linearities either before [14,11] or after feedback [13] and shown that such networks can perform ICA. We will not discuss such work here but only refer the reader to the relevant sources. We will however consider more recent changes to this basic network.

14.3   Independence in Unions of Sources

We first introduce an extension to the above network which can uncover the underlying causes of a data set. Unlike the ICA model, the underlying causes are mixed by a union operator rather than a summation. Thus

$$x = \bigcup_{s \in S} x_s$$
$$= \sum_{x \in S}\left(x_s - \sum_{t < s, t \in S} x_t \otimes x_s\right) \tag{14.3}$$

where x_s is the input from source s and $\otimes$ is defined as the pairwise AND operator. This model may be thought to be more typical of the mixtures of visual components than the ICA model (an object is more liable to be occluded by another than linearly mixed with it) but the underlying task remains the same - find the independent components.

The standard data set used to illustrate this task is the bars data set illustrated in figure 14.2. Each data presentation is a union of bars and the network's task is to identify individual bars [6,26,28,15].

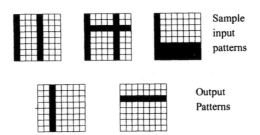

Fig. 14.2. The top line shows sample data (input values) presented to the network. The second layer shows the independent sources which we hope will be the network's response. If this is so, it has clearly identified the suspicious coincidences of lines.

We sometimes describe such a task as a search for " factorial codes": we have many different symbols representing different parts of the environment. These different parts of the environment all occur independently of one another and so the occurrence of a particular input is simply the product of probabilities of the individual code symbols. Thus if x is a sample input, $P(x) = \prod_i P(s_i)$ where $P(x)$ is the probability of the input x and $P(s_i)$ is the probability of the ith factor being present in the inputs.

Using a neural net to create a factorial code means that each neuron should identify one particular source. Such a coding should be deterministic and invertible: each possible input, x, should be guaranteed to cause the same response in the network; also, if we know the code we should be able to go to the environment and identify precisely the input which caused the code reaction from the network.

14.3.1   Factor Analysis

A standard method of finding independent sources of this type is the statistical technique of factor analysis (FA). PCA and FA are closely related statistical techniques both of which achieve an efficient compression of the data but in different manners. They can both be described as methods to explain the data set in a smaller number of dimensions but FA is based on assumptions about the nature of the underlying data whereas PCA is model free.

We can also view PCA as an attempt to find a transformation from the data set to a compressed code, whereas in FA we try to find the linear transformation which takes us from a set of hidden factors to the data set. Since PCA is model free, we make no assumptions about the form of the data's covariance matrix. However FA begins with a specific model which is usually constrained by our prior knowledge or assumptions about the data set. The general FA model can be described by the following relationship:

$$\mathbf{x} = L\mathbf{f} + \mathbf{u} \tag{14.4}$$

where $\mathbf{x}$ is a vector representative of the data set, $\mathbf{f}$ is the vector of factors, L is the matrix of factor loadings and $\mathbf{u}$ is the vector of specific (unique) factors.

We now view the output vector, $\mathbf{y}$, in the negative feedback network as the vector of factors. With FA in mind, we now constrain the model in the previous section so that our model fits the data set we wish to model. In particular we will constrain the model in such a way as to increase its biological plausibility. The network discussed above is clearly implausible at two points:

1. The weights in our network may change from positive to negative (from excitatory to inhibitory) whereas real synapses are either inhibitory or excitatory but cannot change from one to the other (Dale's Law).
2. We allow negative outputs. What would a neuron firing -1 mean?

Because of this, we impose additional constraints on the network rules above: we do not allow a weight to become negative and we do not allow negative outputs. In each case an offending value is simply set to zero. In practice, we have previously shown[10,5] that, with either of these constraints alone, our network weights converge to identify the independent sources of the above data set exactly. We have recently shown that this constraint may be used in general with PCA networks to create FA networks [4] which will identify individual bars in the above data set.

14.3.2   Minimal Overcomplete Bases

Another feature of the model as discussed so far is that it is noise-free, but real neurons operate in an extremely noisy environment. In this section, we show that noise can be extremely beneficial in this problem of identification of the individual sources.

Overcomplete Bases. Artificial neural networks have been shown to be capable of finding optimal bases for data sets in a number of ways. For example, if the basis is of lower dimensionality than the data set, one criterion might be to minimise the mean square distance between any point and its orthogonal projection on the subspace spanned by the basis; in this case, the optimal basis is found by performing PCA of the data set (assuming off-subspace Gaussian residuals). On the other hand we may define optimal to be that basis which identifies clusters most easily in a data set, in which case an exploratory projection pursuit will identify a low order projection in which clusters may be seen.

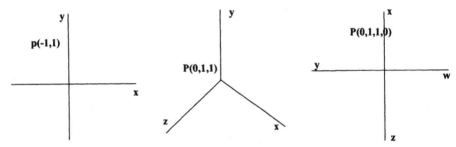

Fig. 14.3. The figure shows how a point P in two dimensional space may be represented by its projection on (a) two basis vectors (b) three basis vectors - an overcomplete basis since only two are actually required to describe the position of any point in this space (c) four basis vectors - this is an overcomplete basis but not a minimum overcomplete basis.

In this section, we will be interested in basis sets which have more basis vectors than the dimensionality of the data set: in figure 14.3, we see an example of a two-dimensional data set which is spanned by two, three and four basis vectors. Our interest is motivated by biological information processing, in particular by the processing of visual information. For such data, we may have an object appear or not but we cannot have the negative of an object appear. We suggest that the coordinates in which the data set is to be described must only have positive elements in the original space if it is to be comprehensible to humans. Thus in figure 14.3, the point P may be described as (-1,1) in the (x,y) - basis, as (0,1,1) in the (x,y,z) -basis and as (0,1,1,0) in the (w,x,y,z)-basis. We are interested in the second and third coordinate systems which only admit positive coordinates.

We have shown how enforcing positive constraints on the coding and position vectors of a space enables humanly-meaningful symbols to be extracted and how a simple change to a PCA artificial neural network enables this to be done using artificial neural networks. We now extend the basic method to show how to find the minimum overcomplete basis (MOB, in figure 14.3 (b) rather than (c)) and what effect this has on the independence finding network.

Let us consider the bars data as an example to illustrate this. Any 1 bar may be formed as a linear combination of the other 15. Thus this data is inherently fifteen-dimensional yet, to be interpretable by humans, we require 16 basis vectors - an overcomplete basis. The previous network used 16 neurons to identify 16 causes. If we have more than 16 neurons, several causes will be split between two or more neurons - bars will be split. In the next section, we show how to achieve the minimal overcomplete basis using no prior knowledge of the number of underlying causes.

Adding Noise. It is well-known [17] that additive noise may be used to introduce a regularisation term into a neural network. The addition of noise into our network acts in a different manner in that it enforces MOB rules.

The above network with rectified output values can be thought of as an example of non-linear PCA, the nonlinearity being the rectification [12]. Non-linear PCA can be shown to be an approximation to the minimisation of $J = E_x(||\mathbf{x} - W\mathbf{y}||^2) = E_x(||\mathbf{x} - W\mathbf{f}(\mathbf{a})||^2)$ where $E_x()$ is the expectation operator. Now we add noise to the outputs, $\mathbf{y}^* = \mathbf{f}(\mathbf{a}) + \mu$ where μ is a vector of independently drawn noise from a zero mean distribution and using $\mathbf{f} = \mathbf{f}(\mathbf{a})$, we define

$$
\begin{aligned}
J_1 &= E_{x,\mu}(||\mathbf{x} - W\mathbf{y}^*||^2) \\
&= E_{x,\mu}(||\mathbf{x} - W(\mathbf{f} + \mu)||^2) \\
&= E_x(||\mathbf{x} - W\mathbf{y}||^2) + E_\mu(||\mu^T W^T W \mu||) \\
&= J + \sum_i ||\mathbf{w}_i||^2 \sigma_i^2
\end{aligned}
\tag{14.5}
$$

where $\mathbf{w}_i$ is the weight into the ith output neuron. We have removed terms containing single expectations of μ from the equation as the noise is drawn from a zero mean distribution. We have also assumed in the last line that the matrix W is orthogonal which is certainly true of the linear version of the network but may not be wholly correct when we use non-linear activation functions. Nevertheless it has been found in practice that the algorithm gives the minimal overcomplete basis.

Intuitively, it can be seen that when the added noise has low magnitude then the first term of equation 14.5 dominates and so (non-linear) PCA is performed in the normal manner. If the noise variance is increased then the learning is moderated by this additional weighted noise term, which has the effect of forcing some weight vectors to have only zero weight values (the degenerate solution).

So the addition of noise to the outputs has the effect of introducing a second pressure into the learning rule of the non-linear PCA algorithm. This is a natural way in which to introduce a sparsification term to the weights. Also as the noise is simply added to the outputs after the application of the non-linearity there is less computational expense than adding a specific weight decay term.

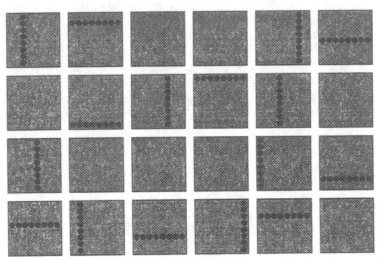

Fig. 14.4. The addition of noise has forced the weights of eight neurons to be zero while the weights of the other 16 have captured the independent components of the data set.

When we use noise on output neurons with the above learning rule on the bars data, we find that e.g. 16 output neurons find non-zero weights and converge to identify a bar each while the other eight output neurons' weights go to zero (figure 14.4).

14.4   Canonical Correlation Analysis

Canonical correlation analysis (CCA) [20] is used when we have two data sets which we believe have some underlying correlation. Consider two sets of input data, from which we draw iid samples to form a pair of input vectors, x_1 and x_2. Then, in classical CCA, we attempt to find the linear combination of the variables which gives us maximum correlation between the combinations. Let

$$y_1 = \mathbf{w}_1\mathbf{x}_1 = \sum_j w_{1j}x_{1j} \tag{14.6}$$

$$y_2 = \mathbf{w}_2\mathbf{x}_2 = \sum_j w_{2j}x_{2j} \tag{14.7}$$

We are therefore trying to maximise

$$J = E\{(y_1y_2) + \frac{1}{2}\lambda_1(1 - y_1^2) + \frac{1}{2}\lambda_2(1 - y_2^2)\}$$

where the λ_i were motivated by the method of Lagrange multipliers to constrain the weights to finite values. We now use the derivative of this function with respect to both the weights, $\mathbf{w}_1$ and $\mathbf{w}_2$, and the Lagrange multipliers, λ_1 and λ_2 to derive learning rules for both:

$$\Delta w_{1j} = \eta x_{1j}(y_2 - \lambda_1 y_1)$$
$$\Delta \lambda_1 = \eta_0(1 - y_1^2)$$
$$\Delta w_{2j} = \eta x_{2j}(y_1 - \lambda_2 y_2)$$
$$\Delta \lambda_2 = \eta_0(1 - y_2^2) \tag{14.8}$$

where w_{1j} is the jth element of weight vector, $\mathbf{w}_1$ etc. The weight update

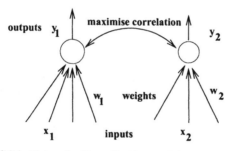

Fig. 14.5. The CCA Network. By adjusting weights, w_1 and w_2, we maximise correlation between y_1 and y_2

rules use a mixture of Hebbian and anti-Hebbian learning. The CCA network presented here may be criticised as a model of biological information processing in that it appears as a non-local implementation of Hebbian learning, i.e. the $\mathbf{w}_1$ weights use the magnitude of y_2 as well as y_1 to self-organise. One possibility is to postulate non-learning connections which join y_2 to y_1 thus providing the information that $\mathbf{w}_1$ requires for learning. Alternatively we may describe the λ_1 parameter as a lateral weight from y_2 to y_1 and so the learning rules become

$$\Delta w_{1j} = (\eta \lambda_1) x_{1j} \left(\frac{y_2}{\lambda_1} - y_1 \right)$$

$$\Delta \lambda_1 = \eta_0 (1 - y_1^2)$$

where we have had to incorporate a λ_1 term into the learning rate. Perhaps the second of these suggestions is more plausible than the first as a solution to the non-local feature of the previous learning rules since non-learning connections hardwire some activation passing into the cortex. We have used this network [18] to extract depth information from a data set which is an abstraction [1] of random dot stereograms.

We have previously shown [18] that nonlinear correlations may be extracted by a network using

$$y_1 = \sum_j w_{1j} \tanh(v_{1j} x_{1j}) = \mathbf{w}_1 \mathbf{f}_1 \text{ and} \qquad (14.9)$$

$$y_2 = \sum_j w_{2j} \tanh(v_{2j} x_{2j}) = \mathbf{w}_2 \mathbf{f}_2 \qquad (14.10)$$

and derived the learning rules

$$\Delta \mathbf{w}_1 = \eta \mathbf{f}_1 (y_2 - \lambda_1 y_1)$$

$$\Delta \mathbf{v}_{1i} = \eta \mathbf{x}_{1i} \mathbf{w}_{1i} (y_2 - \lambda_1 y_1)(1 - \mathbf{f}_1^2)$$

and similarly for the second neuron. These rules have been shown to be more effective than the linear neurons when the correlations between inputs is non-linear.

14.4.1   Extracting Multiple Correlations

We have previously used competition between pairs of output neurons (each y_1 matched with a y_2 neuron) to channel the learning of each pair of neurons. In this chapter, we use a somewhat different method which is more satisfactory with respect to mathematical exposition but less satisfying as a BIANN. The competition method gives equally good results when applied to our test data.

Now we wish to extract pairs of maximal correlations and ensure that all pairs are orthogonal to each other.

Thus the criterion we use is

$$J = E\{(\mathbf{y}_1^T \mathbf{y}_2) + \Lambda_1 (I - \mathbf{y}_1 \mathbf{y}_1^T)\} + \Lambda_2 (I - \mathbf{y}_2 \mathbf{y}_2^T) \qquad (14.11)$$

with $\Lambda_i, i = 1, 2$ now a matrix of Lagrange multipliers. This gives us the learning rules

$$\Delta \mathbf{w}_1 = \eta \mathbf{f}_1 (y_2 - \Lambda_1 y_1)$$

$$\Delta \mathbf{v}_1 = \eta \mathbf{w}_1 (y_2 - \Lambda_1 y_1) \mathbf{x}_1 (1 - \mathbf{f}_1^2)$$

$$\Delta \Lambda_1 = \eta_0 (I - \mathbf{y}_1 \mathbf{y}_1^T)$$

where

$$\mathbf{y}_1 = W_1^T \mathbf{x}_1 \qquad\qquad (14.12)$$

with W_1 the $M \times N$ matrix of weights connecting $\mathbf{x}_1$ to $\mathbf{y}_1$. Similarly with W_2, Λ_2.

This model may be viewed as an abstraction of two data streams (e.g. sight and sound) identifying an entity in the environment by identifying the maximal non-linear correlations between the data streams. An alternative use of this type of method is to attempt to identify independent components of data streams by minimising the correlations between data sets.

14.4.2   Using Minimum Correlations to Extract Independent Sources

Just as a principal component analysis network can be changed into a minor component analysis network by changing the sign of the learning rate, we can change our CCA network into a network which searches for that linear combination of the data set which has the minimum mutual correlation by changing the sign of the learning.

However one difficulty with this is that when performing descent on the correlations, it is possible to create negative correlations. We therefore require to add an additional constraint that ensures that $E(y_1 y_2) \geq 0$. For the non-linear neurons, this gives us the learning rule

$$\Delta \mathbf{w}_1 = -\eta \mathbf{f}_1 ((I + \Lambda_3)\mathbf{y}_2 - \Lambda_1 \mathbf{y}_1)$$
$$\Delta \mathbf{v}_1 = -\eta \mathbf{w}_1 (I + \Lambda_3)(\mathbf{y}_2 - \Lambda_1 \mathbf{y}_1)\mathbf{x}_1 (1 - \mathbf{f}_1^2)$$
$$\Delta \Lambda_1 = -\eta_0 (I - \mathbf{y}_1 \mathbf{y}_1^T)$$
$$\Delta \Lambda_3 = \eta_2 \mathbf{y}_2 * \mathbf{y}_1'$$

where we have used the Λ_i to denote matrices of parameters. Similarly with $\mathbf{w}_2$, $\mathbf{v}_2$ and Λ_2. This method then ensures that we do not find weight pairs which give a negative non-linear correlation.

We wish to emulate biological information processing in that e.g. auditory information cannot choose which pathway it uses. All early processing is done by the same neurons. In other words, these neurons see the information at each time instant. We therefore structure our data set so that

$$\mathbf{x}_1 = \{m_{1,1}(t), m_{1,2}(t), ...m_{1,1}(t+1), m_{1,2}(t+1), ...m_{1,2}(t+P_1)\}$$
$$\mathbf{x}_2 = \{m_{2,1}(t), m_{2,2}(t), ...m_{2,1}(t+1), m_{2,2}(t+1), ...m_{2,2}(t+P_2)\}$$

Good extraction of single sinusoids from a mixture of sinusoids (e.g. for 2×2 square mixes $\mathbf{m}_1, \mathbf{m}_2$) have been obtained using only

$$\mathbf{x}_1 = \{m_{1,1}(t), m_{1,2}(t), m_{1,1}(t+P_1), m_{1,2}(t+P_1)\}$$
$$\mathbf{x}_2 = \{m_{2,1}(t), m_{2,2}(t), m_{2,1}(t+P_2), m_{2,2}(t+P_2)\}$$

where the delays P_1 and P_2 need not be equal.

Both methods have been successfully used on extraction of sinusoids; we report some experimental work in the next section.

14.4.3   Experiments

We have previously [18] investigated the general problem of maximising correlations between two data sets when there may be an underlying non-linear relationship between the data sets: we generate artificial data according to the prescription:

$$x_{11} = 1 - \sin\theta + \mu_1 \tag{14.13}$$
$$x_{12} = cos\theta + \mu_2 \tag{14.14}$$
$$x_{21} = \theta + \mu_3 \tag{14.15}$$
$$x_{22} = \theta + \mu_4 \tag{14.16}$$

where θ is drawn from a uniform distribution in $[-\pi, \pi]$ and $\mu_i, i = 1, ..., 4$ are drawn from the zero mean Gaussian distribution $N(0, 0.1)$. Equations 14.13 and 14.14 define a circular manifold in the two dimensional input space while equations 14.15 and 14.16 define a linear manifold within the input space where each manifold is only approximate due to the presence of noise $(\mu_i, i = 1, ..., 4)$.

Thus $\mathbf{x}_1 = \{x_{11}, x_{12}\}$ lies on or near the circular manifold $x_{11}^2 + x_{12}^2 = 1$ while $\mathbf{x}_2 = \{x_{21}, x_{22}\}$ lies on or near the line $x_{21} = x_{22}$.

With this data set, the linear network achieved a maximal correlation of 0.623 while the non-linear network achieved a correlation of 0.865. We have achieved equally good results with higher dimensional spheres and planes.

Now we embed signals in non-linear spaces, i.e. instead of using random numbers for the θ values, we sample from signals. So for example, using a sphere and a plane to contain the signals we might have

$$\mathbf{x}_{11} = \sin(s_1) * \cos(s_2) \tag{14.17}$$
$$\mathbf{x}_{12} = \sin(s_1) * \sin(s_2) \tag{14.18}$$
$$\mathbf{x}_{13} = \cos(s_1) \tag{14.19}$$
$$\mathbf{x}_{21} = s_1 \tag{14.20}$$
$$\mathbf{x}_{22} = s_2 \tag{14.21}$$

where e.g. $s_1(t) = \sin(t/8) * \pi, s_2(t) = \sin(t/5 + 3) * \pi$ for each of $t = 1, 2, ..., 500$. Thus we have embedded two sine waves on the surface of a sphere and on the plane. With this data set, we can easily recover single sines from mixtures. Similar results have been achieved when $\mathbf{x}_2$ is a linear mixture of the sines though the quality of the recovered signal deteriorates when both $\mathbf{x}_1$ and $\mathbf{x}_2$ are non-linear mixtures. Speech signals have so far only provided limited success.

14.5 ϵ-Insensitive Hebbian Learning

We can use the residuals equation 14.2 after feedback to define a general cost function associated with this network as

$$J = f_1(\mathbf{e}) = f_1(\mathbf{x} - W\mathbf{y}) \qquad (14.22)$$

where in the above $f_1 = ||.||^2$, the (squared) Euclidean norm. It is well known (e.g. [27,2]) that with this choice of $f_1()$ the cost function is minimised with respect to any set of samples from the data set on the assumption of Gaussian noise on the samples.

It can be shown that, in general (e.g. [27]), the maximisation of J is equivalent to minimising the negative log probability of the error or residual, $\mathbf{e}$, which may be thought of as the noise in the data set. Thus if we know the probability density function of the residuals, we may use this knowledge to determine the optimal cost function.

To motivate this discussion, consider the bars data set of section 14.3. There is inherent noise within this data set caused by every horizontal bar occluding every vertical bar and vice versa. This noise is rather more kurtotic than Gaussian (Table 14.1 gives an example).

Table 14.1. Approximate probabilities that the given number of pixels will be firing in a given bar. Probabilities for four to seven pixels not shown are less than 0.05.

Number of Pixels	0	1	2	3	8
Probability	0.3	0.34	0.17	0.05	0.125

An approximation to these density functions is the (one-dimensional) function

$$p(e) = \frac{1}{2 + \epsilon} \exp(-|e|_\epsilon) \qquad (14.23)$$

where

$$|e|_\epsilon = \begin{cases} 0 & \forall |e| < \epsilon \\ |e - \epsilon| & \text{otherwise} \end{cases} \qquad (14.24)$$

with ϵ being a small scalar ≥ 0. Using this model of the noise, the optimal $f_1()$ function (to minimise the negative log probability of the error) is the ϵ - insensitive cost function

$$f_1(e) = |e|_\epsilon \qquad (14.25)$$

Therefore when we use this function in the (non-linear) negative feedback network we get the learning rule

$$\Delta W \propto -\frac{\partial J}{\partial W} = -\frac{\partial f_1(e)}{\partial e}\frac{\partial e}{\partial W}$$

which gives us the learning rules

$$\Delta w_{ij} = \{ \begin{matrix} 0 & \text{if } |x_j - \sum_k w_{kj}y_k| < \epsilon \\ \eta.y_i.sign(x_j - \sum_k w_{kj}y_k) = \eta.y.sign(e) & \text{otherwise} \end{matrix}$$

(14.26)

where $sign(t) = 1$ if $t > 0$ and $sign(t) = -1$ if $t < 0$.

We see that this is a simplification of the usual Hebbian rule using only the sign of the residual rather than the residual itself in the learning rule. We will find that in the linear case (section 14.3) allowing ϵ to be zero gives generally as accurate results as non-zero values but the data sets in section 14.4 require non-zero ϵ because of the nature of their innate noise.

14.5.1   Is this a Hebbian Rule?

The immediate question to be answered is does this learning rule qualify as a Hebbian learning rule given that the term has a specific connotation in the artificial neural networks literature? We may consider e.g. covariance learning [19] to be a different form of Hebbian learning but at least it still has the familiar product of inputs and outputs as a central learning term.

The first answer to the question is to compare what Hebb wrote with the equations above: *"When an axon of cell A is near enough to excite a cell B and repeatedly or persistently takes part in firing it, some growth process or metabolic change takes place in one or both cells such that A's efficiency, as one of the cells firing B, is increased."*

We see that Hebb is quite open about whether the pre-synaptic or the post-synaptic neuron would change (or both) and how the mechanism would work. Indeed it appears that Hebb considered it unlikely that his conjecture could ever be verified (or falsified) since it was so indefinite [7]. Secondly, it is not intended that this method replaces traditional Hebbian learning but that it coexists as a second form of Hebbian learning. This is biologically plausible - *"Just as there are many ways of implementing a Hebbian learning algorithm theoretically, nature may have more than one way of designing a Hebbian synapse"*[3]. Indeed the suggestion that long term potentiation *"seems to involve a heterogeneous family of synaptic changes with dissociable time courses"* seems to favour the coexistence of multiple Hebbian learning mechanisms which learn at different speeds.

The new rule is very simple and equally very much a BIANN.

14.5.2   Extraction of Sinusoids

In this section we consider the standard ICA problem discussed elsewhere in this volume. The non-linear extension to Oja's subspace algorithm [17,14]

$$\Delta w_{ij} = \eta(x_j f(y_i) - f(y_i) \sum_k w_{kj} f(y_k)) \qquad (14.27)$$

has been used to separate signals into their significant subsignals. As a first example, we repeat Karhunen and Joutsensalo's [17] experiment to separate samples of a sum of sinusoids into their component parts. The experimental data consists of samples of a signal composed of a sum of sinusoids in noise:

$$x(t) = \sum_j A_j cos(2\pi f_j t - \theta_j) + \omega_t \qquad (14.28)$$

The amplitudes, A_j, frequencies, f_j, and phases θ_j, are all unknown. We use white noise, $\omega_t \sim N(0, 0.05)$ where t denotes time. The input vector is a vector comprising a randomly drawn instance of $x(t)$ and that of the 14 subsequent times, $t + 1, ..t + 14$.

We use the ϵ-insensitive Hebbian rules with the tanh() non-linearity in equation 14.1. An example of a recovered signal of a three sinusoid experiment (with $f_j = 0.15, 0.2$ and 0.25) is shown in figure 14.6.

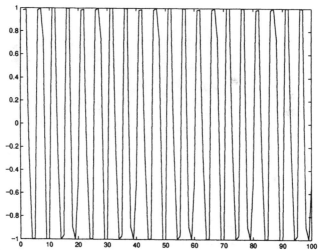

Fig. 14.6. One recovered signal from a mixture of three sinusoids, $\epsilon = 0.5$.

In all cases the recovered signals have the correct frequencies. It is our finding that any non-zero value of ϵ gives these types of results but that $\epsilon = 0$ fails to completely remove one sinusoid from the other two.

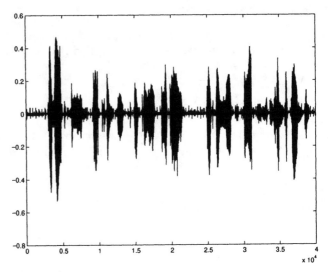

Fig. 14.7. The original signal.

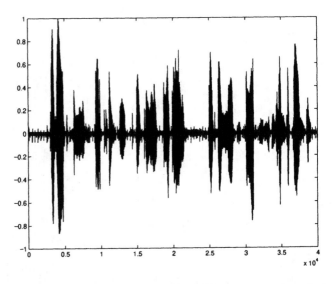

Fig. 14.8. The recovered signal using the minor ϵ-insensitive component analysis.

Table 14.2. The relative power of each signal in each of the three mixes. Signal 3 is very much the weakest yet is readily recovered.

	Signal 1	Signal 2	Signal 3
Mixture 1	$2.8 * 10^{11}$	$0.3 * 10^{11}$	1
Mixture 2	$1.1 * 10^{11}$	$0.2 * 10^{11}$	1
Mixture 3	$2.5 * 10^{11}$	$1.1 * 10^{11}$	1

Perhaps the most frequent use of ICA is in the extraction of independent voices from a mixture of voices. Now it has often been noted that individual voices are kurtotic[16] and it has been shown that the nonlinear PCA network with nonlinearities matching the signal's statistics can readily extract individual voices from a mixture. The ϵ-insensitive rule performs similarly; in the next section we report on a slightly different method for extraction of a voice in specific circumstances.

14.5.3   Noise Reduction

Rather than maximising equation 14.22, we may consider its minimisation which results in those eigenvectors of the data covariance matrix which have the smallest eigenvalues. One of the most interesting application of ϵ-insensitive PCA is in the finding of the smaller magnitude components. Xu *et al* [29] have shown that by changing the sign in the PCA learning rule we may find the minor components of a data set - those eigenvectors which have the smallest associated eigenvalues. We use this technique to extract a signal from a mixture of signals: we use three speech signals and mix them so that the power of the weakest signal constitutes a small fraction of the total power in the mixes. Table 14.2 gives an example of the relative power of each of the signals in each of the three mixtures. Figure 14.7 shows the weakest power signal while Figure 14.8 shows the recovered output - the low power signal has been recovered.

We emphasise that we are using a linear network (and hence the second order statistics of the data set) to extract the low power voice. The network's rules are

$$y_i = \sum_{j=1}^{N} w_{ij} x_j$$

$$x_j(t+1) \leftarrow x_j(t) - \sum_{k=1}^{M} w_{kj} y_k$$

$$\Delta w_{ij} = \begin{cases} 0 & \text{if } |x_j(t+1)| < \epsilon \\ -\eta.y.sign(x_j(t+1)) & \text{otherwise} \end{cases}$$

The mixture is such that humans cannot hear the third voice at all in any of the three mixtures and yet the MCA method reliably finds the signal with least power. The original third signal (figure 14.7) and its recovered form (Figure 14.8) when we use $\epsilon = 0$, learning rate $= 0.1$ should be compared.

We note that a limitation of this method might seem to be that the signal to be recovered must be swamped by the noise: if there is a component of the noise which is lower power than the signal and also more kurtotic than the mixture this will be recovered. However there is one situation (and perhaps a frequent one given the symmetry of our ears and surroundings) in which MCA may be useful: if the mixing matrix is ill-conditioned. Consider the mixing matrix $A = \begin{pmatrix} 0.8 & 0.4 \\ 0.9 & 0.4 \end{pmatrix}$ whose determinant is -0.04. Its eigenvalues are 1.2325 and -0.0325. The major principal component will be the term which is constant across the two signals and so the smaller component will be the residual when this has been removed. Two voice signals were mixed using this matrix and one recovered by the MCA method is shown in figure 14.9 which should be compared with the original signal shown in figure 14.7. Most interesting is that the more ill-conditioned the mixing matrix, the better is the recovery of the lesser amplitude voice.

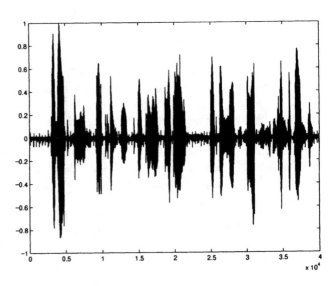

Fig. 14.9. The recovered signal from MCA when the mixing matrix is ill-conditioned.

14.6   Conclusion

In this chapter, we have made a plea for keeping to artificial neural networks which remain close to biological models. We recognise that such models are never perfect abstractions of reality but believe that we have shown that models may be both biologically plausible and reliable independence-seekers.

We have reviewed three specific models:

1. The first network is related to factor analysis and is used to identify the independent causes from visual data sets (where the image is a union rather than a sum of underlying causes).
2. The second network is derived to perform canonical correlation analyis and nonlinear versions are shown to extract underlying signals from nonlinear mixtures.
3. The third network is shown to be optimal with respect to non-Gaussian noises and is shown to solve the cocktail party problem; a specific use related to minor component analysis extracts signals which are submerged in very high amplitude noise.

All three of these models are the subject of continuing research (as are others from the same stable). We are continuing to expand the breadth of problems tackled by these models using e.g. kernel methods to extend the basic methods.

References

1. S. Becker. Mutual information maximisation: Models of cortical self-organization. *Network: Computation in Neural Systems*, 7:7–31, 1996.
2. C. Bishop. *Neural Networks for Pattern Recognition*. Clarendon Press, Oxford, 1995.
3. T. H. Brown and S. Chatterji. Hebbian Synaptic Plasticity, *The Handbook of Brain Theory and Neural Networks*, MIT Press, 1995.
4. D. Charles and C. Fyfe. Discovering independent sources with an adapted PCA network. In *Proceedings of The Second International Conference on Soft Computing, SOCO97*, Sept. 1997.
5. D. Charles and C. Fyfe. Modelling multiple cause structure using rectification constraints. *Network: Computation in Neural Systems*, 1998.
6. P. Dayan and R. S. Zemel. Competition and multiple cause models. *Neural Computation*, 7:565–579, 1995.
7. Y. Fregnac. Hebbian Synaptic Plasticity: Comparative and Developmental Aspects. *The Handbook of Brain Theory and Neural Networks*, MIT Press, 1995.
8. C. Fyfe. PCA properties of interneurons. In *From Neurobiology to Real World Computing, ICANN 93*, 183–188, 1993.
9. C. Fyfe. Introducing asymmetry into interneuron learning. *Neural Computation*, 7(6):1167–1181, 1995.
10. C. Fyfe. A neural net for pca and beyond. *Neural Processing Letters*, 6(1):33–41, 1997.

11. C. Fyfe and R. Baddeley. Non-linear data structure extraction using simple hebbian networks. *Biological Cybernetics*, **72**(6):533–541, 1995.

12. C. Fyfe and D. Charles. Using noise to form a minimal overcomplete basis. In *Seventh International Conference on Artificial Neural Networks, ICANN99*, 1999.

13. M. Girolami and C. Fyfe. An extended exploratory projection pursuit network with linear and non-linear lateral connections applied to the cocktail party problem. *Neural Networks*, 1997.

14. M. Girolami and C. Fyfe. Stochastic ica contrast maximisation using oja's nonlinear pca algorithm. *International Journal of Neural Systems*, **8**(5):661–678, 1997.

15. G.E. Hinton, P. Dayan, and M. Revow. Modelling the manifolds of images of handwritten digits. *IEEE Transactions on Neural Networks*, **8**(1):65–74, 1997.

16. C. Jutten and J. Herault. Blind separation of sources,part 1: An adaptive algorithm based on neuromimetic architecture. *Signal Processing*, **24**: 1–10, 1991.

17. Juha Karhunen and Jyrki Joutsensalo. Representation and separation of signals using nonlinear pca type learning. *Neural Networks*, **7**(1):113–127, 1994.

18. P.L.Lai and C. Fyfe. A neural network implementation of canonical correlation analysis. *Neural Networks*, **12**:1391–1397, 1999.

19. R. Linsker. From basic network principles to neural architecture. *Proceedings of National Academy of Sciences*, 1986.

20. K. V. Mardia, J.T. Kent, and J.M. Bibby. *Multivariate Analysis*. Academic Press, 1979.

21. E. Oja. A simplified neuron model as a principal component analyser. *Journal of Mathematical Biology*, **16**:267–273, 1982.

22. E. Oja. Neural networks, principal components and subspaces. *International Journal of Neural Systems*, **1**:61–68, 1989.

23. E. Oja, H. Ogawa, and J. Wangviwattana. Pca in fully parallel neural networks. In Aleksander & Taylor, editor, *Artificial Neural Networks,2*, 1992.

24. B. D. Ripley. *Pattern Recognition and Neural Networks*. Cambridge University Press, 1996.

25. T.D. Sanger. Analysis of the two-dimensional receptive fields learned by the generalized hebbian algorithm in response to random input. *Biological Cybernetics*, 1990.

26. E. Saund. A multiple cause mixture model for unsupervised learning. *Neural Computation*, **7**:51–71, 1995.

27. A. J. Smola and B. Scholkopf. A tutorial on support vector regression. Technical Report NC2-TR-1998-030, NeuroCOLT2 Technical Report Series, 1998.

28. R. H. White. Competitive Hebbian learning: Algorithm and demonstration. *Neural Networks*, **5**:261–275, 1992.

29. L. Xu, E. Oja, and C. Y. Suen. Modified Hebbian learning for curve and surface fitting. *Neural Networks*, **5**:441–457, 1992.

Index